I0100307

Review of Atlantic and Eastern Pacific Anthiine Fishes

Frontispiece. *Baldwinella vivanus.* Gulf of Mexico; live male. Photographed by D. Ross Robertson at the Charlotte Aquarium.

Review of Atlantic and Eastern Pacific Anthiine Fishes (Teleostei: Perciformes: Serranidae), with Descriptions of Two New Genera

William D. Anderson, Jr.
Grice Marine Biological Laboratory
College of Charleston
Charleston, South Carolina

Phillip C. Heemstra
South African Institute for Aquatic Biodiversity
Republic of South Africa

American Philosophical Society
Philadelphia • 2012

TRANSACTIONS
of the
American Philosophical Society
Held at Philadelphia
for Promoting Useful Knowledge
Volume 102, Part 2

Copyright © 2012 by the American Philosophical Society for its *Transactions* series.
All rights reserved.

ISBN: 978-1-60618-022-8
US ISSN: 0065-9746

Library of Congress Cataloguing-in-Publication Data

Anderson, William D. (William Dewey), 1933-
  Review of Atlantic and Eastern Pacific anthiine fishes (Teleostei: Perciformes: Serranidae),
with descriptions of two new genera / William D. Anderson, Jr., and Phillip C. Heemstra.
    p. cm.—(Transactions of the American Philosophical Society; v. 102, pt. 2)
  Includes bibliographical references and index.
  ISBN 978-1-60618-022-8   1. Serranidae—Classification.   I. Heemstra, Phillip C.   II. Title.
  QL638.S48A63 2012
  597'.736—dc23

                                                                              2012003334

Front Cover. Small group of *Pronotogrammus martinicensis*. West Flower Garden Bank, northwestern Gulf of Mexico (depth: 86 meters). Photograph courtesy of Emma Hickerson and Lance Horn, National Oceanic and Atmospheric Administration (Flower Garden Banks National Marine Sanctuary/National Undersea Research Center, University of North Carolina at Wilmington).

Back Cover and Frontispiece. *Baldwinella vivanus.*

William D. Anderson, Jr.: Grice Marine Biological Laboratory, College of Charleston, 205 Fort Johnson, Charleston, South Carolina 29412-9110, USA; e-mail andersonwd@cofc.edu

Phillip C. Heemstra: South African Institute for Aquatic Biodiversity, Private Bag 1015, Grahamstown 6140, Republic of South Africa; e-mail p.heemstra@saiab.ac.za

Dedicated to our wives, Barbara Anderson and Elaine Heemstra, for their long years of support and encouragement.

I observed some fine anthiae, which belong to the order of lutjans, a fish held sacred by the Greeks, who attributed to them the power of hunting the marine monsters from waters they frequented. Their name signifies flower, *and they justify their appellation by their shaded colours, their shades comprising the whole gamut of reds, from the paleness of the rose to the brightness of the ruby, and the fugitive tints that clouded their dorsal fin.*

Jules Verne, 1869 (1963:175–176), **20,000 Leagues Under the Sea**

TABLE OF CONTENTS

LIST OF FIGURES

LIST OF TABLES

LIST OF MAPS

PREFACE

Thirty-seven species of Anthiinae in 15 genera (two of them new) are considered herein. Twenty-three of those species are residents of the Atlantic Ocean, and 14 dwell in the eastern Pacific. The genus *Choranthias* is erected for the sister species *salmopunctatus* and *tenuis*, heretofore placed in the genus *Anthias*, and the genus *Baldwinella* is erected for three closely related species, *Hemanthias aureorubens, H. vivanus,* and *Pronotogrammus eos.* Keys to the genera and species are presented, and accounts are provided for 46 of the 52 taxa included (the exceptions are *Acanthistius* and the five species considered in the key to that genus).

ACKNOWLEDGMENTS

The following assisted us in numerous ways, including providing space to work, donating specimens, sending specimens on loan, examining specimens, supplying information on specimens, and furnishing literature, illustrations, and radiographs: C. Allué, K. Amaoka, R. Arrindell, C. C. Baldwin, M. L. Bauchot, the late E. B. Böhlke, the late J. E. Böhlke, M. Bougaardt, R. Britz, B. Brown, L. H. Bullock, G. H. Burgess, W. A. Bussing, P. Campbell, A. Carvalho Filho, D. Catania, T. Cekalovic K., D. Clark, F. C. Coleman, L. J. V. Compagno, M. T. Craig, O. Crimmen, M. Desoutter, W. N. Eschmeyer, M. N. Feinberg, C. R. Gilbert, J. Gilhen, A. C. Gill, R. G. Gilmore, Jr., M. F. Gomon, K. M. Green, A. S. Harold, K. E. Hartel, W. F. Hoffman, P. A. Hulley, S. Jewett, G. D. Johnson, D. Jones, J.-C. Joyeux, the late G. Krefft, A. L. Labadie, M. Lamboeuf, R. J. Lavenberg, K. R. Luckenbill, O. J. Luiz Jr., P. P. Maier, B. M. Martin, J. E. McCosker, M. McGrouther, the late J. F. McKinney, D. R. McMillan, R. Meléndez C., N. A. Menezes, N. R. Merrett, S. Meyers, D. W. Nelson, M. Nuttall, L. Palmer, N. V. Parin, B. Ranchod, J. E. Randall, S. J. Raredon, W. J. Richards, C. D. Roberts, C. R. Robins, R. H. Robins, M. A. Rogers, W. A. Roumillat, M. H. Sabaj, W. G. Saul, J. A. Seigel, C. L. Smith, D. G. Smith, W. F. Smith-Vaniz, V. G. Springer, W. A. Stahl, B. W. Stender, R. E. Stobbs, A. Y. Suzumoto, M. S. Taylor, G. C. Van Dyke, H. J. Walker, Jr., M. W. Westneat, the late A. Wheeler, J. T. Williams, the late L. P. Woods, J. E. Wright, D. Wyanski, F. A. Young, and B. Zgliczynski.

William A. Roumillat examined histological sections of gonads and provided us with his expert opinion on the sexuality of several anthiine species. Barbara S. Anderson, the first author's wife, helped compile some of the data. William N. Eschmeyer was a continual source of information on the names of fishes and on the rules of zoological nomenclature. Improvements or verifications of translations were provided by W. D. Anderson, III (German), J. Boylan, J. Escobar, and N. J. Salcedo (Spanish), M. Salustiano de Castro (Portuguese), and N. A. Chamberlain (French). Gabe Sataloff made the maps.

The following provided color photographs of specimens: K. Amaoka, C. C. Baldwin, L. H. Bullock, G. H. Burgess, E. D'Antoni, D. E. De Vore, J. K. Dooley, A. Edwards, S. Elmore, D. D. Flescher, D. Golani, C. B. Grimes, E. Hickerson, L. Horn, G. de Quadros Benvegnú Lé, O. J. Luiz Jr., R. Martin, N. V. Parin, J. E. Randall, D. R. Robertson, S. W. Ross, W. F. Smith-Vaniz, F. F. Snelson, the late B. A. Thompson, C. A. Wenner, and B. Zgliczynski. Antony S. Harold, D. Hurlbert, S. Money, D. Polack, J. L. Russo, N. J. Salcedo, and A. E. Sanders furnished photographic assistance. Carole C. Baldwin, E. D'Antoni, and J.-E. Trecartin prepared the drawings. Kunio Amaoka, D. E. De Vore, A. Edwards, D. D. Flescher, P. Louisy, O. J. Luiz Jr., N. V. Parin, J. E. Randall, D. R. Robertson, and S. W. Ross allowed us to reproduce color images. Allen Press Publishing Services, the American Society of Ichthyologists and Herpetologists, the Florida Fish and Wildlife Conservation Commission, the Food and Agriculture Organization of the United Nations, Wiley-Blackwell, and R. H. Kuiter granted permission to use published illustrations. Michelle S. Brew and Sarah F. Goldman helped with computer applications. Katie Anweiler helped prepare the index, and Gail A. Meyers assisted in the final assembly of this review. The late J. F. McKinney called WDA's attention to the passage on anthiines in Jules Verne's *20,000 Leagues Under the Sea*.

The first author is indebted to the Foundation for Research Development, Republic of South Africa, for the award of a research fellowship that allowed him to work at the J. L. B. Smith Institute of Ichthyology (now South African Institute for Aquatic Biodiversity, Grahamstown) from May through July 1992; the writing of the manuscript for this contribution was initiated during the tenure of that fellowship. A portion of the research reported herein was conducted during WDA's appointment in October 2002 as a Short-Term Visiting Scientist at the National Museum of Natural History, Smithsonian Institution. At the invitation of K. E. Carpenter, WDA participated in the workshop on the living marine resources of the eastern central Atlantic that was sponsored by the Food and Agriculture Organization of the United Nations and Centro Oceanográfico de Canarias and held in July 2004 at Santa Cruz de Tenerife, Canary Islands; at that workshop he examined for the first time the two known specimens of *Meganthias carpenteri*.

Both authors owe thanks to the late Frederick H. Berry who taught us much about fishes and showed us that life is not all work. Over a period of many years, Albert E. Sanders has been a frequent source of assistance and wise counsel to the first author—very importantly, suggesting that we submit the manuscript for this review to the American Philosophical Society. Carole C. Baldwin and her interns (Cristina Castillo, Donald Griswold, and Brendan Luther) tested the generic key, G. David Johnson commented on a short section of the manuscript, and Baldwin and William F. Smith-Vaniz read a presubmission draft of this document and offered numerous helpful suggestions. Mary McDonald, Editor and Director of Publications, American Philosophical Society, and Heather Willison, Senior Project Editor, S4Carlisle Publishing Services, provided advice and guided the manuscript through the publication process. This is contribution number 382 of the Grice Marine Biological Laboratory, College of Charleston.

INTRODUCTION

Years ago we began collaborating on studies of the Atlantic and eastern Pacific species of the serranid subfamily Anthiinae. Only a limited part of our results has been published (Anderson and Heemstra, 1980; Heemstra and Anderson, 1983; Anderson and Heemstra, 1989). In addition, WDA's interest has led to nine other papers on anthiines from those waters: two on eastern Atlantic species (Anderson, 2006; inpress), two on western Atlantic species (Anderson, 2003; Anderson and García-Moliner, 2012), and five on eastern Pacific species (Anderson et al., 1990; Anderson and Randall, 1991; Anderson and Baldwin, 2000, 2002; Anderson, 2008). In this work we present our efforts to date, recognizing 15 genera, two described herein as new, and 37 species of Atlantic and eastern Pacific Anthiinae. Because the classification of the Anthiinae is largely unsatisfactory, it was necessary to make several changes in allocation of species to genera in order to reflect current information.

Katayama and Amaoka (1986) restricted *Anthias* to include only Atlantic forms, removing Indo-Pacific species more appropriately regarded as representatives of *Pseudanthias*, *Franzia*, or *Mirolabrichthys*. Although not clearly stated, Katayama and Amaoka (1986:217–219, 221) apparently considered *Anthias* to include the following species: *anthias*, *asperilinguis*, *helenensis*, *menezesi*, *nicholsi*, *salmopunctatus*, *tenuis*, and *woodsi*. We modify their concept of the genus *Anthias* to exclude *salmopunctatus* and *tenuis*, which, as shown by Baldwin (1990), appear to be sister species and warrant placement in a genus distinct from *Anthias*. In addition, we include in *Anthias* both *noeli*, a species described as new by Anderson and Baldwin (2000), and *cyprinoides*, formerly assigned to *Holanthias* by Katayama and Amaoka (1986). Following Baldwin (1990), we consider *Holanthias martinicensis* and *Pronotogrammus multifasciatus* as congeneric and the only members of the genus *Pronotogrammus*; in addition we erect a new genus (*Choranthias*) for *Anthias salmopunctatus* and *A. tenuis* and a new genus (*Baldwinella*) for *Hemanthias aureorubens*, *Hemanthias vivanus*, and *Pronotogrammus eos*.

There are about 215 valid described species of Anthiinae, classified in at least 29 genera (including the two described herein); in addition, there are a number of other known undescribed species of Anthiinae, and almost certainly others await recognition or collection. Remarkably, the original descriptions of more than half of the valid species of the subfamily have been published since 1971—48 of them in the years 1979–1982. This dramatic increase in the number of anthiine species known is in large part due to more intensive collecting efforts over the last 40 years; in particular, use of ichthyocides and SCUBA have allowed sampling at previously inaccessible depths. During this period, one ichthyologist, John E. Randall, stands out for his contributions to our knowledge of the Anthiinae; he has made many collections of those fishes and has described or codescribed, as new, more than 60 species of this subfamily.

The Anthiinae include small to medium size fishes that inhabit tropical to temperate seas worldwide in shallow to moderate depths, usually on rocky bottoms or coral reefs. These fishes are usually brightly colored, and most feed on zooplankton near the bottom, to which individuals quickly escape at the approach of predators. All anthiines (with the exception of *Epinephelides armatus*, see Moore et al., 2007; and probably *Lepidoperca aurantia*, see Roberts, 1989) for which the sexuality has been unambiguously determined are protogynous hermaphrodites. Many anthiines display sexual dichromatism, and some are sexually dimorphic, particularly in fin structure. Anthiines often occur in aggregations, with males typically attending large harems (Anderson et al., 1990).

Lindquist and Clavijo (1994) provided an estimate of the combined population densities of *Holanthias martinicensis* (= *Pronotogrammus martinicensis*) and *Hemanthias vivanus* (= *Baldwinella vivanus*) observed off North Carolina as 188,000 to 325,000 per 0.5 km². Their estimate was based on point counts, visual transects, and video transects made during a dive of a submersible in August 1987 over a continental shelf-edge reef about 95 kilometers southeast of Cape Fear (33°12.8'N, 77°18.0'W) in 96 to 109 meters of water. Gilmore (R. Grant Gilmore, Jr., *in litt.* to WDA, 23 August 1996) related that in the Western Hemisphere, anthiines are the numerically dominant fish species "in the 100–300 m transition between the upper slope/wall fauna of the euphotic zone and the deeper epibathyal faunas." Gilmore's statement was based on observations he has made from submersibles over a number of years. In a study of fishes associated with shelf-edge hard bottoms off North Carolina, Quattrini and Ross (2006:145–146) mentioned that anthiines were the most abundant fishes (>49%) seen on most hard bottom habitats. They collected four species of anthiines (*Anthias tenuis* [= *Choranthias tenuis*], *Hemanthias leptus*, *H. vivanus* [= *Baldwinella vivanus*], and *Pronotogrammus martinicensis*), often from mixed schools, with a suction sampler on the Johnson-Sea-Link submersible (Harbor Branch Oceanographic Institution).

Although the Atlantic and eastern Pacific anthiine fauna is represented by an appreciable number of species, many of which are quite common, there is remarkably little general literature to which one can refer for reliable identifications. Fitch (1982) and Heemstra (1995) provided keys to and accounts for four eastern Pacific species, and Anderson and Baldwin (2000) furnished keys to the species of *Anthias* and of all species of anthiines then known from the eastern Pacific. Anderson (2003) presented accounts for the species then known to occur in the western central Atlantic, a key to the species found in the eastern Atlantic (Anderson, 2006), and accounts for the species inhabiting the eastern central Atlantic (Anderson, in press).

ABBREVIATIONS AND METHODS

Institutional abbreviations are as listed in Leviton et al. (1985), except for the following additions: BAMZ (Bermuda Aquarium, Museum & Zoo), FSBC (Florida Fish and Wildlife Conservation Commission, Fish and Wildlife Research Institute, St. Petersburg), HBOM (Harbor Branch Oceanographic Museum, Fort Pierce, Florida), IIPB (Instituto de Ciencias del Mar, Barcelona, Spain), NMFS (U.S. National Marine Fisheries Service), NMSZ (National Museums of Scotland, Edinburgh), SAIAB (formerly RUSI; South African Institute for Aquatic Biodiversity, Grahamstown, Republic of South Africa), SU (Stanford University collections, now at California Academy of Sciences), UF (Florida Museum of Natural History, University of Florida, Gainesville), UFES (Laboratório de Ictiologia, Universidade Federal do Espírito Santo, Vitória, ES, Brazil), and ZIN (Zoological Institute, Russian Academy of Sciences, St. Petersburg). Other abbreviations are: ICZN—International Code of Zoological Nomenclature (International Commission on Zoological Nomenclature), IUCN—International Union for Conservation of Nature, SL—standard length, TL—total length, NL—notochord length (distance from snout to posterior end of notochord, measurement used for larvae before notochord flexion), ROV—Remotely Operated Underwater Vehicle.

In each generic account, characters in the generic diagnosis form part of the generic description and are not repeated unless necessary for clarification. Similarly, in each species account, characters in the appropriate generic and species diagnoses and generic description form part of the species description and are not repeated unless necessary for clarification. The types of ctenoid scales found in anthiines are illustrated in Fig. 1, and the configurations of supraneural bones, anterior neural spines, and anterior dorsal pterygiophores occurring in Atlantic and eastern Pacific anthiines are shown in Fig. 2. Frequency distributions for counts of soft rays in the dorsal and anal fins, rays in the pectoral fin, total numbers of gillrakers on the first gill arch, and scales in the lateral line and around the caudal peduncle for all species except those of *Acanthistius* are in Tables 2 to 7.

The maps show the positions of capture for material we examined for which latitudes and longitudes were available and for collections where reasonably accurate localities could be determined from other information associated with the specimens studied. In Map 10, in addition to localities from which we examined specimens of *Pronotogrammus martinicensis*, we plotted sites of collection reported by Coleman, 1982, that were unique to her study. The position plotted for *Holanthias caudalis* is the locality given in the original description (Trunov, 1976). In some instances symbols on maps represent more than one collection.

We used Eschmeyer's electronic version of the *Catalog of Fishes* to check dates of publication, authorships, spellings, and related items—see: http://research .calacademy.org/ichthyology/catalog/fishcatmain.asp and consulted Bauchot et al. (1984) for information on anthiine types in the MNHN. Other methods used are those of Anderson and Heemstra (1980), as modified and clarified by Anderson et al. (1990) and Anderson and Baldwin (2000). Some of those methods are reiterated or clarified below. Osteological data are from radiographs and/or cleared and stained material. Vertebral counts are presented as: total number (number of precaudal + number of caudal), e.g., 26 (10 + 16). We follow Mabee (1988) in using "supraneural" instead of "predorsal" to refer to the series of bones anterior to the pterygiophores supporting the dorsal fin, and Johnson and Patterson (1993:557) and Patterson and Johnson (1995) in using the term "epineurals" for the intermuscular bones that traditionally have been called "epipleurals" in perciform fishes. The notation used in the formulae for configurations of supraneural bones, anterior neural spines, and anterior dorsal pterygiophores is that of Ahlstrom et al. (1976:297): e.g., 0/0+0/2/1+1/1/ where each zero (0) represents a supraneural bone, each slash (virgule) a neural spine, and the numerals the number of spines supported by the first and subsequent pterygiophores in a secondary association (see Fig. 2).

Two types of ctenoid scales occur in adult anthiines; one type has both marginal cteni and ctenial bases (remains of old cteni) in the posterior field (see Hughes, 1981) (Fig. 1A); the other type has marginal cteni but lacks ctenial bases in the posterior field (Fig. 1B). Roberts (1993) called the first type of scale transforming ctenoid and the latter type peripheral ctenoid. Herein we refer to scales (i.e., those along the midlateral aspect of the body) as lacking or possessing ctenial bases in the posterior field. (Roberts, 1993, presented a comparative study of spined scales in the Teleostei that is a good introduction to the morphological variation found in the ctenoid scales of anthiines.) In addition to ctenoid scales, one species included herein, *Trachypoma macracanthus*, has mostly cycloid scales in larger individuals. Also, a few species have secondary squamation (accessory scales or squamulae) at bases of head and body scales.

Counts of lateral-line scales are of tubed scales and were made on either side, depending on condition of the specimen. Counts of caudal-peduncle scales are of the number of scales around the narrowest part of the peduncle (this count is difficult to make on some specimens). Counts of gillrakers are of those on the first gill arch, include all rudiments, and were made on the right side where possible. Sums of tubed lateral-line scales plus total number of gillrakers on the first gill arch are based on single counts of each character from each specimen examined.

Measurements were made with needle-point dial calipers to nearest 0.1 mm and are presented as percentages of standard length, except internarial distance is given as the quotient of the snout length divided by the distance between the nares (on one side of a specimen). Measurements from the anterior end of the snout were taken from the premaxillary symphysis; those involving the orbit (snout length, orbit diameter) were of the bony orbit. Measurement of the orbit was of horizontal diameter. Depth of body was measured from dorsal-fin origin vertically to ventral midline of body. Anal-fin length was from origin of fin to distal end of longest ray with fin depressed. Pelvic-fin length was of longer (either left or right) fin. Lengths of caudal-fin lobes were taken from the middle of fin base to tips of longest rays.

ANTHIINAE

Johnson (1983) differentiated the Serranidae from the Percichthyidae (*sensu* Gosline, 1966) on the basis of three reductive specializations (absence of the posterior uroneural, procurrent spur, and third preural radial cartilages), and demonstrated that serranids share at least one innovative specialization (the presence of three spines on the opercle); these specializations support the hypothesis that the family is monophyletic. Johnson (1983, 1988), following Gosline (1966), recognized three subfamilies: the Serraninae, Epinephelinae, and Anthiinae, but identified a synapomorphy for only the epinepheline species. Two characters (one reductive, the other innovative) that may prove useful in defining the Anthiinae are absence of a tooth plate on the second epibranchial (discussed by Baldwin, 1990, and interpreted by her as a synapomorphy for anthiines) and vertebral number. Anderson and Heemstra (1989) and Anderson et al. (1990) discussed vertebral number in the Serranidae, and Anderson and Heemstra (1989) considered counts of 24 or 25 total vertebrae as the most primitive state in the Serranidae and 26, 27, and 28 as progressively more derived states. Anderson et al. (1990) accepted, for the time being, the absence of the second epibranchial tooth plate and high vertebral number (26 to 28, usually 26) as autapomorphies delimiting the subfamily Anthiinae.

Our examination of a 159-mm SL cleared and stained specimen of *Serranus cabrilla* (Linnaeus, 1758), the type species of *Serranus*, which is the type genus of the subfamily Serraninae, causes us to question the utility of the absence of the second epibranchial tooth plate as a defining character for the Anthiinae. On this specimen (collected off Kasouka, Eastern Cape Province, South Africa), we found that the second epibranchial lacks a tooth plate; there is, however, a tooth plate firmly attached to the second infrapharyngobranchial that has its posterior end adjacent to, but unattached to, the anterior end of the second epibranchial.

"Based on morphology of gill arches, configuration of dorsal-fin pterygiophores and number of vertebrae, Baldwin . . . proposed that *Acanthistius* and *Trachypoma*, formerly treated as a serranine and epinepheline, respectively, may be cladistically primitive anthiines" (Baldwin and Johnson, 1993:242). *Acanthistius* Gill, 1862, is an enigma. We have found the second epibranchial tooth plate to be present on specimens of four of the six species studied, *A. patachonicus* (Jenyns, 1840), *A. pictus* (Tschudi, 1846), *A. sebastoides* (Castelnau, 1861), and *A. serratus* (Cuvier, 1828), apparently present on the single individual of *A. fuscus* Regan, 1913a, examined, and missing from the only specimen of *A. brasilianus* (Cuvier, 1828) examined. To complicate matters further, *Acanthistius* has 26 vertebrae. Johnson (1983) acknowledged 26 as the vertebral number for *Acanthistius*, but stated, without explanation, that it is a serranine. The relationships of *Acanthistius* will be treated in a future paper by P. C. Heemstra; consequently, except for including the genus and Atlantic and eastern Pacific species in keys and in Table 1, we do not consider those taxa further.

Trachypoma macracanthus Günther, 1859, is widely distributed in the South Pacific; in the eastern Pacific it has been reported from Easter Island and from the Desventuradas Islands off Chile (Yáñez-Arancibia, 1975; Randall and Cea Egaña, 1984; Pequeño and Lamilla, 1996a, b, and 2000; Rojas and Pequeño, 1998a; Randall et al., 2005). In view of its apparent affinities with the Anthiinae (see preceding paragraph), we consider *T. macracanthus* to be an anthiine and provide accounts of it and of the genus *Trachypoma*.

The taxonomic position of the Serranidae, long considered a member of the Order Perciformes, Suborder Percoidei, has recently received renewed attention. Smith and Craig (2007), using molecular data from 180 species of Acanthomorpha, arrived at novel conclusions about the taxonomy of Perciformes, Percoidei, Trachinoidei, and Serranidae. With two exceptions they moved the members of the traditional Serranidae from the Percoidei to the Scorpaenoidei. They recognized five clades within the traditional Serranidae, advocating, among other things, elevation of the groupers (minus *Niphon*) to the familial level as the Epinephelidae and retaining within their revised Serranidae the Serraninae and Anthiinae, essentially intact, while adding the Trachininae (= former Trachinidae) to that family. They removed *Acanthistius* from the Anthiinae, based on their data from *Acanthistius ocellatus*, considering that species to be *incertae sedis* among the Percoidei (see Craig and Hastings, 2007:15, who considered *A. ocellatus* to be *incertae sedis* among the Serranidae), and moved *Zalanthias kelloggi* (= *Plectranthias kelloggi*) from the Anthiinae to the Serraninae. Wiley and Johnson (2010), using morphological synapomorphies, proposed a Linnaean classification of teleosts based on monophyletic groups, placing the Serranidae in the Order Scorpaeniformes, Suborder Serranoidei (see Imamura and Yabe, 2002).

Mooi and Gill (2010a) cautioned against the acceptance of phylogenies, such as the one proposed by Smith and Craig (2007), in which character-based cladistics has been abandoned for trees derived through the use of statistical algorithms, leading to the failure to identify or recognize synapomorphies, the *sine qua non* of phylogenetic systematics. In the COMMENTS section of *Copeia* 2010, No. 3, Chakrabarty (2010) responded to an earlier contribution by Mooi and Gill (2008) in which they had expressed the concern that there is "a crisis in systematics," a concern amplified later in Mooi and Gill (2010a). Further on in the COMMENTS section, Mooi and Gill (2010b) took the opportunity to restate and clarify their position, and in the last paper in that section Smith (2010) replied to Mooi and Gill in some detail, ending his essay with "As I see it, Mooi and Gill's (2010) impending crisis is, in reality, the impending phylogenetic and taxonomic renaissance" (p. 523). Although not discounting the possibility that molecular biology holds great potential for elucidating teleost relationships, Wiley and Johnson (2010:125) seem reluctant to jump on the bandwagon: "Molecular data are of increasing importance in unraveling the evolutionary relationships of teleost fishes. . . . However, molecular work is still in its infancy. Studies with different genes tend to yield different results that are difficult to compare, because the taxa are frequently different."

Major Groups of Anthiinae

Anderson et al. (1990:927) concluded, based in large part on data provided by Johnson (1984), that number of principal caudal-fin rays (branched rays + 2), number of supraneural bones, and type of ctenoid scale "may be helpful in clarifying the generic classification of the Anthiinae." Based on available information (Johnson, 1984; Anderson et al., 1990; Baldwin, 1990), it appears that among the Anthiinae presence of 17 principal caudal-fin rays is the primitive condition—15 the derived, presence of three supraneural bones is primitive—one or two derived, and presence of ctenial bases

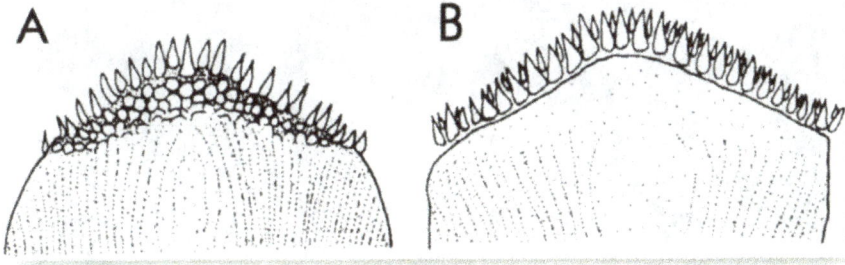

Figure 1. Scales of anthiine fishes (posterior fields of scales toward top of page). **A.** Ctenoid scale with ctenial bases present in posterior field. **B.** Ctenoid scale with only marginal cteni (i.e., no ctenial bases in the posterior field). Drawings by Carole C. Baldwin.

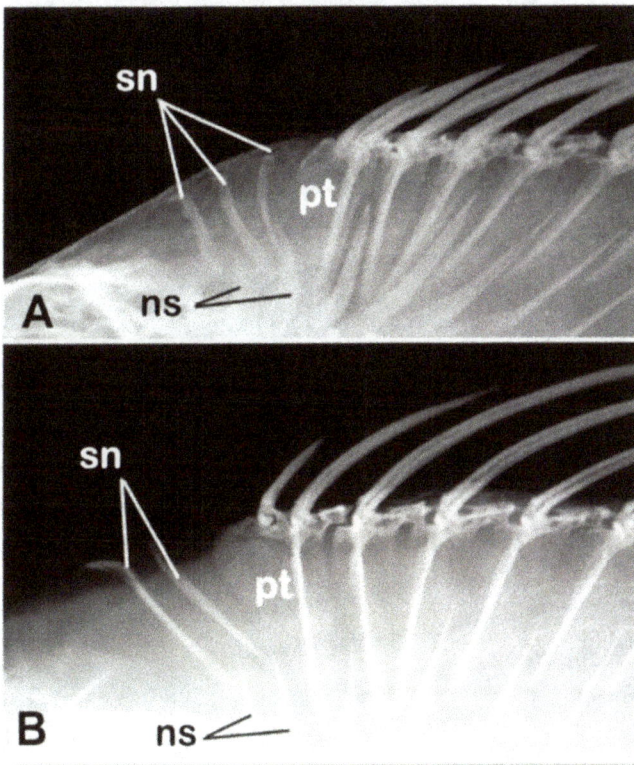

Figure 2. Configurations of supraneural bones, anterior neural spines, and anterior dorsal pterygiophores in anthiine fishes; **ns** = neural spine, **pt** = dorsal pterygiophore, **sn** = supraneural bone. In legends for **A** and **B** below, each zero (0) represents a supraneural bone, each slash (virgule) a neural spine, and the numerals the number of spines supported by the first and subsequent pterygiophores in a secondary association.

　　A. 0/0+0/2/1+1/1/, presumably the more primitive condition; from radiograph of specimen of *Hypoplectrodes semicinctum*; IZUC 2768, 151 mm SL.

　　B. 0/0/2/1+1/1/, apparently the derived condition; from radiograph of specimen of *Anthias anthias*; IIPB 673/1981, 167 mm SL. Scans of radiographs by Sean Money; figure assembled by A. E. Sanders.

Figure 3. *Trachypoma macracanthus*. Easter Island, eastern South Pacific; BPBM 6613, 156 mm SL. Photograph by John E. Randall.

Figure 4. *Lepidoperca coatsii*. Nightingale Island, Tristan da Cunha Group, central South Atlantic; SAIAB 31523, 110 mm SL. Photograph by T.G. Andrew; published as plate 1.H in Andrew et al. (1995:42). Courtesy of South African Institute for Aquatic Biodiversity.

Figure 5. *Caprodon longimanus*. Easter Island, eastern South Pacific; BPBM 6644, 276 mm SL.
Photograph by John E. Randall.

Figure 6. *Hypoplectrodes semicinctum*. Juan Fernández Islands, eastern South Pacific.
Photograph by Alvaro Sepúlveda; courtesy of Rudie H. Kuiter; published in Kuiter (2004:116).

Figure 7. *Plectranthias exsul.* Juan Fernández Islands, eastern South Pacific; holotype, ANSP 127843, 158 mm SL. Photograph by Amanda L. Labadie. ©Amanda L. Labadie, Academy of Natural Sciences of Philadelphia.

Figure 8. *Plectranthias garrupellus.* Off east coast of Florida, western North Atlantic; ca. 50 mm SL. Photograph by R. Grant Gilmore; published in Bullock and Smith (1991); reproduced courtesy of Florida Fish and Wildlife Conservation Commission.

Figure 9. *Plectranthias nazcae.* Nazca Ridge, eastern South Pacific. From a color transparency received from N. V. Parin; published in Anderson (2008), *Proceedings of the Biological Society of Washington*; reproduced by permission of Allen Press Publishing Services. ©2008 Biological Society of Washington.

Figure 10. *Plectranthias parini*. Sala y Gómez Ridge, eastern South Pacific; holotype, USNM 312925, 85 mm SL. Drawing by Jo-Ellen Trecartin; pattern of coloration drawn from color transparency, courtesy of N. V. Parin (erratum: posterior bar does not extend as far out on dorsal fin as shown).

Figure 11. *Hemanthias leptus*. Off Louisiana, Gulf of Mexico; FSBC 12023, male, 313 mm SL. Photograph by Lewis H. Bullock; published in Bullock and Smith (1991); reproduced courtesy of Florida Fish and Wildlife Conservation Commission.

Figure 12. *Hemanthias peruanus*. Gulf of Chiriqui, Panama, eastern North Pacific. Photograph by D. Ross Robertson.

Figure 13. *Hemanthias signifer*. Gulf of Chiriqui, Panama, eastern North Pacific. Photograph by D. Ross Robertson.

Figure 14. *Choranthias salmopunctatus*. Saint Paul's Rocks, Mid-Atlantic Ridge, equatorial Atlantic. Photograph by João L. Gasparini, courtesy of Osmar J. Luiz Jr. Figure 1 in Luiz et al. (2007), *Journal of Fish Biology* 70:1284; published by Wiley-Blackwell.

Figure 15. *Choranthias tenuis.* Off Curaçao, southern Caribbean Sea; USNM 406397, 85 mm SL. Photograph by D. Ross Robertson and Carole C. Baldwin.

Figure 16. *Anthias anthias.* Off Cataluña, Spain, Mediterranean Sea; length 140 mm. Photograph by Patrick Louisy; courtesy of Rudie H. Kuiter; published in Kuiter, 2004:16. ©Patrick Louisy.

Figure 17. *Anthias asperilinguis.* South America, Atlantic Coast; holotype, BMNH 1974.10.4.1, 143 mm SL. Published in Boulenger (1895).

Figure 18. *Anthias cyprinoides*. Southwest of Pagalu (Annobón) Island, eastern South Atlantic; paratype, HUMZ 100018, 222 mm SL. From color transparency received from Kunio Amaoka.

Figure 19. *Anthias helenensis*. North of Saint Helena Island, eastern South Atlantic; holotype, HUMZ 100156, 163 mm SL. From color transparency received from Kunio Amaoka.

Figure 20. *Anthias menezesi*. Off southern Brazil, western South Atlantic; holotype, MZUSP 11765, 132 mm SL. Photograph by Joseph L. Russo; published in Anderson and Heemstra (1980); courtesy of American Society of Ichthyologists and Herpetologists.

Figure 21. *Anthias nicholsi.* Off Virginia, western North Atlantic; GMBL 81-146, 88 mm SL. Photograph by Donald D. Flescher.

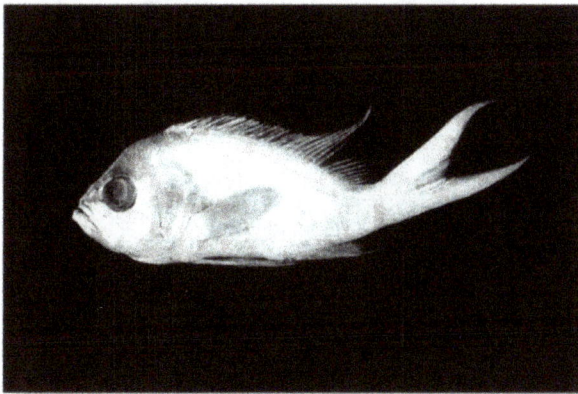

Figure 22. *Anthias noeli.* Seamount southeast of Isla San Cristobal, Galápagos Islands, eastern Pacific; holotype, USNM 353113, male, 167 mm SL. Photograph by Donald Hurlbert.

Figure 23. *Anthias woodsi.* Off South Carolina, western North Atlantic; UF 101423, 232 mm SL. Photograph by Steve W. Ross.

Figure 24. *Holanthias fronticinctus*. Off Saint Helena Island, eastern South Atlantic; male, ca. 210 mm SL. Photograph by Alasdair Edwards.

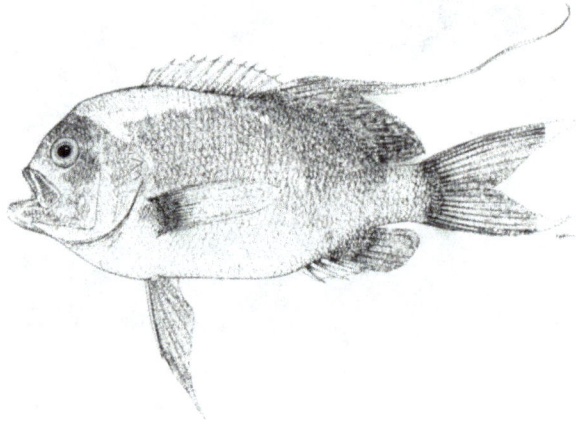

Figure 25. *Meganthias carpenteri*. Off Nigeria, eastern Atlantic; holotype, USNM 386079, male, 301 mm SL. Drawing by Emanuela D'Antoni. ©Food and Agriculture Organization of the United Nations.

Figure 26. *Odontanthias hensleyi*. Off west coast of Puerto Rico, Mona Passage, western North Atlantic; holotype, USNM 400888, male, 155 mm SL. Photograph by Denise E. De Vore.

Figure 27. *Pronotogrammus martinicensis*. Off west coast of Florida, Gulf of Mexico; FSBC 18002, 73 mm SL. Photograph by Lewis H. Bullock; published in Bullock and Smith (1991); reproduced courtesy of Florida Fish and Wildlife Conservation Commission.

Figure 28. *Pronotogrammus multifasciatus*. Gulf of Chiriqui, Panama, eastern North Pacific. Photograph by D. Ross Robertson.

Figure 29. *Baldwinella aureorubens*. Western North Atlantic, 242 mm fork length. Photograph by Donald D. Flescher; courtesy of Rudie H. Kuiter; published in Kuiter (2004:19).

Figure 30. *Baldwinella eos*. Gulf of Chiriqui, Panama, eastern North Pacific. Photograph by D. Ross Robertson.

Figure 31. *Baldwinella vivanus*. Western Atlantic, 130 mm fork length. Photograph by Donald D. Flescher; courtesy of Rudie H. Kuiter; published in Kuiter (2004:18).

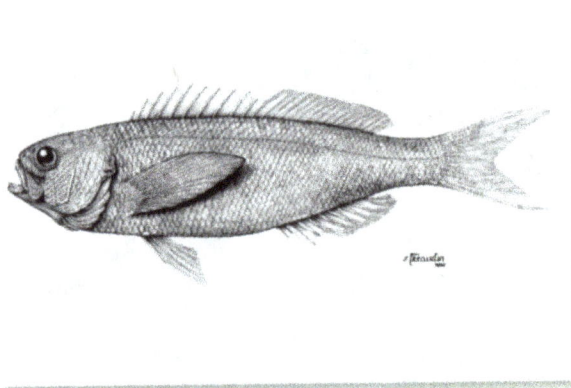

Figure 32. *Anatolanthias apiomycter*. Near southwest end of Nazca Ridge, eastern South Pacific; holotype, USNM 309202, 94 mm SL. Drawing by Jo-Ellen Trecartin.

in the posterior fields of scales is primitive—absence of ctenial bases in the posterior fields derived. Anderson et al. (1990:928) wrote:

> Among anthiines there is a strong correlation in number of principal caudal-fin rays, number of predorsal [= supraneural] bones, and type of ctenoid scale. Species with 17 principal rays and three predorsal bones usually have scales in which ctenial bases have been retained in the posterior field, but among those with 15 principle [sic] rays apparently all lack ctenial bases in the posterior field (Anderson, unpublished data). Although all three of the presumed derived states (15 principal caudal-fin rays, one or two predorsal bones, and absence of ctenial bases in the posterior field) are reductive, the shared possession of all three may be indicative of propinquity of descent. (Based on our incomplete data we speculate that the sequence of appearance of these derived characters in the main line of anthiine evolution was: loss of ctenial bases in the posterior field, reduction in number of principal caudal-fin rays, and reduction in number of predorsal bones.)

Atlantic and eastern Pacific genera of Anthiinae can be separated into two sections that seem to represent distinct groups, one assemblage (*Acanthistius, Trachypoma, Lepidoperca, Caprodon, Hypoplectrodes,* and *Plectranthias*—in part) possessing the three traits presumed to be primitive and the other (*Hemanthias, Choranthias, Anthias, Holanthias, Meganthias, Odontanthias, Pronotogrammus, Baldwinella,* and *Anatolanthias*) with the alternative derived states. *Plectranthias garrupellus* occupies an intermediate position in this scheme, having one of the derived character states (absence of ctenial bases in the posterior fields of scales) and two of the primitive states.

Characters Shared by Anthiinae Treated Herein

To reduce repetition and to provide a ready reference for the morphology of the species, we list the following characters shared by all Atlantic and eastern Pacific species of Anthiinae: posterior margin of bony opercle with three spinous processes, middle one best developed; branchiostegal rays 7; pelvic-fin rays I, 5; pleural ribs usually present only on vertebrae 3 through 10, rarely present on 11th vertebra; no spur on posteriormost ventral procurrent caudal-fin ray (see Johnson, 1975); penultimate ventral procurrent caudal-fin ray not shortened basally; parhypural with well-developed hypurapophysis; autogenous hypurals 5—no hypural fusions; epurals 3; uroneurals 1 pair (posterior pair absent).

Otoliths

Fitch (1982) championed the utility of otoliths for the identification of stomach contents in advanced stages of digestion because otoliths are frequently found in the guts of piscivorous species and are often identifiable to genus or species. He described and illustrated (p. 7, fig. 5) the sagittal otoliths of four eastern Pacific anthiines, noting that "in the eastern North Pacific, anthiins [sic] apparently are a choice prey for many predators" (p. 6). Fitch mentioned that the otoliths of the four eastern Pacific anthiines that he studied are commonly found in the scats of sea lions (*Zalophus californianus*) that haul out on Islotes Island, north of La Paz, in the Gulf of California. We know of no other published studies on the otoliths of eastern Pacific Anthiinae and are aware of none on the Atlantic species.

Maxillary Hook

A shelf or rostrally directed hook on the labial border of the distal end of the maxilla (see Anderson et al., 1990:926, fig. 2) is sometimes present in anthiines. The hook is variously developed to absent in those species in which it is found. We have observed it in a number of anthiines, including *Anatolanthias apiomycter*, *Anthias anthias*, *A. asperilinguis*, *A. nicholsi*, *Baldwinella eos*, *B. vivanus*, *Choranthias salmopunctatus*, *C. tenuis*, *Hemanthias leptus*, *H. peruanus*, *H. signifer*, *Odontanthias hensleyi*, *Pronotogrammus martinicensis*, and *Sacura parva*. The maxillary hook is easily overlooked because it is usually hidden by the upper lip. The function of the hook, if indeed there is one, is obscure.

Sexuality

Sadovy de Mitcheson and Liu (2008), in an extensive consideration of hermaphroditism in teleosts, emphasized the importance of distinguishing between functional and nonfunctional hermaphroditism. In nonfunctional hermaphroditism (p. 4):

> A proportion of individuals of a species or a population may exhibit both testicular and ovarian tissues but only . . . reproduce as either male or female. . . . From a functional perspective, which ignores gonadal morphology, this state is gonochoristic. A species or population is considered to exhibit functional hermaphroditism if a proportion of individuals functions as both sexes at some time during their lives.

Sadovy de Mitcheson and Liu (2008:17) asserted that determining functional sexuality in serranids requires careful study, including the histological examination of a broad size range of juveniles and adults. In order to confirm protogyny, it is necessary to find "gonads in transition between female and male function, together with, if possible, the observation of sex change under field or laboratory conditions."

Thresher (1984) summarized the information (current through about 1981) on the reproductive biology of anthiines. As noted by Anderson and Baldwin (2000) protogyny has been reported in species of a number of anthiine genera, including *Anthias* (*A. anthias*—Reinboth, 1964), *Baldwinella* (*B. vivanus* [as *Hemanthias vivanus*]—Hastings, 1981), *Hemanthias* (*H. peruanus*—Coleman, 1983), *Hypoplectrodes* (*H. huntii* [as *Ellerkeldia huntii*] and *H. maccullochi*— Jones, 1980, and Webb and Kingsford, 1992, respectively), *Pronotogrammus* (*P. martinicensis* [as *Holanthias martinicensis*]—Coleman, 1981), *Pseudanthias* (*P. squamipinnis* [as *Anthias squamipinnis*]—Fishelson, 1970, and Shapiro in several studies on the behavioral aspects of sex reversal: e.g., Shapiro, 1986), and *Sacura* (*S. margaritacea*—Reinboth, 1963). Also, Robins and Starck (1961) stated that *Plectranthias garrupellus* is probably protogynous, Heemstra (1973) provided evidence for protogyny in *Pseudanthias conspicuus* (as *Anthias conspicuus*), and based on their studies Anderson and Baldwin (2000:382) felt "that it is reasonable to conclude" that both *Anthias nicholsi* and *A. noeli* are protogynous. Herein we provide evidence for protogyny in *Anthias woodsi*.

It might be assumed, based on the preceding paragraph, that all anthiines are protogynous and monandric. Nevertheless, in a study of the Serranidae of the eastern Gulf of Mexico, Bullock and Smith (1991) presented evidence for protogyny in the anthiine *Hemanthias leptus*, but, based on finding a ripening male of 86 mm SL (p. 21, fig. 8B) and a sexually mature male of 61 mm SL (p. 207, pl. ID), considered the possibility that *H. leptus* is diandric, noting, however, that additional work is needed to confirm this. Roberts (1989:584), after examining gross morphology and gonadal histology, found "no evidence of protogyny as diagnozed [*sic*] by the criteria of Sadovy and Shapiro (1987)" in the New Zealand anthiine *Lepidoperca aurantia* and stated that "it is reasonable to conclude that . . . [it] is not sexually dimorphic and is probably a gonochorist." Similarities in age distributions of males and females and histological demonstration that the gonads are made up exclusively of either ovarian or testicular tissues convinced Moore et al. (2007) that the southwestern Australian anthiine *Epinephelides armatus* is gonochoristic.

Based on a number of studies, groupers (Serranidae: Epinephelinae), in general, have been thought to display monandric protogyny, but diandry has been confirmed in *Cephalopholis taeniops*, *Epinephelus andersoni*, *Plectropomus leopardus*, and *P. maculatus* (Sadovy de Mitcheson and Liu, 2008). Transformation from male to female has been induced in the laboratory in two groupers (*Cephalopholis boenak* and *Epinephelus akaara*), showing that those species have the capability of bidirectional sex change under experimental conditions (Sadovy de Mitcheson and Liu, 2008), "but there is no evidence that this pattern occurs in wild populations" (Erisman et al., 2009:E84). Sadovy and Colin (1995:961) found that sexuality in *Epinephelus striatus* "is essentially gonochoristic, with potential for sex change" and that "the juveniles pass through a bisexual stage of gonadal development." Rhodes and Sadovy (2002:865) in a study of the reproductive biology of *Epinephelus polyphekadion* were unable to resolve the sexual pattern, but, based on their work and that of others, wrote "that males of this species may have two developmental pathways, directly from the juvenile and through sex change." Erisman et al. (2008) found that population and histological data indicate that *Mycteroperca rosacea* is gonochoristic; "although some individuals pass through a bisexual juvenile phase prior to sexual maturation, adult sex change does not occur" (p. 31). Smith (1971) found evidence suggesting that two species of another epinepheline genus, *Liopropoma*, are secondary gonochorists (i.e., derived from hermaphroditic ancestors).

The remaining serranid subfamily, the Serraninae, displays an amazing variety and flexibility in sexuality, running the gamut from simultaneous hermaphroditism in *Diplectrum*, *Hypoplectrus*, *Serranus* (Smith, 1975; Bubley and Pashuk, 2010), and *Serraniculus* (Sadovy de Mitcheson and Liu, 2008) through protogyny in *Centropristis* (Lavenda, 1949; Wenner et al., 1986), to secondary gonochorism in *Paralabrax clathratus* (Smith and Young, 1966). Hovey and Allen (2000:459) looked at the reproductive patterns of six populations of *Paralabrax maculatofasciatus* and found that they "are best represented by a spectrum ranging between gonochorism and protogynous hermaphroditism." A more recent study of four species of *Paralabrax* (including *P. clathratus* and *P. maculatofasciatus*) showed all to be "functionally gonochoristic, with a non-functional bisexual juvenile phase of gonad development and clear separation of male and female tissues" (Sadovy and Domeier, 2005:130). Sadovy de Mitcheson and Liu (2008:15) cited studies reporting a combination of simultaneous and sequential hermaphroditism in *Serranus baldwini* and *S. fasciatus* (=S. psittacinus)—"smaller individuals within a social group are simultaneous hermaphrodites, while the largest often lose female function and reproduce exclusively as a male." Erisman and Hastings (2011) termed this last sexual pattern androdieocy.

Erisman and Hastings (2011:357) found that their phylogenetic reconstruction of sexual patterns in the Serranidae *sensu stricto* (i. e., the Anthiinae plus Serraninae) "is consistent with the hypothesis that protogyny is the ancestral condition in the family from which other sexual patterns evolved" and that most anthiines and some serranines retained protogyny.

This discussion highlights the shortcomings of making a priori assumptions about the reproductive biology of serranid fishes and points out the necessity of evaluating the sexuality of each species on its own merits.

Feeding Habits

There is an appreciable literature demonstrating a correlation between gillrakers (development and number) and feeding habits in serranid fishes. Rojas et al. (1998:940, table 2) presented a table showing numbers of gillrakers and prey preferences in serranid fishes (including eight anthiines), along with literature references. Their table shows that species with smaller numbers of gillrakers tend to feed on larger prey (shrimps and fishes), whereas those with larger numbers tend to select smaller prey (amphipods, copepods, ostracods, pteropods, and salps). Considering the fact that many anthiines covered in the present study have fairly large numbers of rather long and closely set gillrakers, it is not surprising to find that much of their diets (where known) consists of zooplankton gleaned from the water column.

Species with fewer gillrakers (such as *Acanthistius* spp., *Trachypoma macracanthus*, *Hypoplectrodes semicinctum*, and most species of *Plectranthias*) can be expected to consume larger prey. They are also more sedentary and solitary, being ambush predators that feed on benthic crustaceans and small fishes. These nonschooling species are further characterized in having a truncate or convex caudal fin, typical of bottom-dwelling quick-start ambush predators.

KEY TO THE GENERA OF ATLANTIC AND EASTERN PACIFIC ANTHIINAE

In the key, unless otherwise noted, scale counts are of numbers of tubed lateral-line scales (excluding any posterior to base of caudal fin); gillraker counts are of total numbers on first gill arch (including rudiments, if present). Species names are given if genera are monotypic or represented in the area by a single species. This key may not be useful for extralimital species. In Table 1 some characters useful in distinguishing the genera are presented.

1a. Scales ctenoid with ctenial bases present proximal to marginal cteni (Fig. 1A) or in one genus (*Trachypoma*) scales mostly cycloid but ctenoid scales present on some specimens
.... 2

1b. Scales ctenoid, with only marginal cteni (i.e., no ctenial bases present proximal to marginal cteni; Fig. 1B)... 7

2a. Spines in dorsal fin 11–13; ventral border of preopercle with strong antrorse spines 3

2b. Spines in dorsal fin 10; ventral border of preopercle without antrorse spines, except in *Hypoplectrodes semicinctum* and in some *Plectranthias* 4

3a. Soft rays in dorsal fin 14–18; soft rays in anal fin 7–10; scales ctenoid .... *Acanthistius*

3b. Soft rays in dorsal fin 13; soft rays in anal fin 6; scales mainly cycloid, but some specimens with numerous ctenoid scales—mainly posteriorly (individuals less than about 50 mm SL may have only ctenoid scales) .... *Trachypoma macracanthus*

4a. Soft rays in dorsal fin 19–22; scales 48–65 5

4b. Soft rays in dorsal fin 15–19 (usually 15–17); scales 28–51 (usually 28–48) 6

5a. Maxilla without scales; preopercle with 2 or 3 antrorse spines; gillrakers 17–20; scales 48–55 (usually 49–51) .... *Hypoplectrodes semicinctum*

5b. Maxilla with scales; preopercle serrate, but without antrorse spines; gillrakers 34–37; scales 58–65, usually 61–64 (counts only from eastern Pacific specimens) ....
.... *Caprodon longimanus*

6a. Gillrakers 26–31; maxilla without scales (occasionally with a few scales in *P. exsul* and *P. nazcae*) .... *Plectranthias*

6b. Gillrakers 31–36; maxilla with scales .... *Lepidoperca coatsii*

7a. Maxilla naked 8

7b. Maxilla with scales 10

8a. Gillrakers 14–18; scales 28 or 29; ventral border of preopercle with 1–3 antrorse spines
.... *Plectranthias*

8b. Gillrakers 31–43; scales 36–71; no antrorse spines on preopercle....................... 9

9a. Circum-caudal-peduncular scales 22–29; gillrakers 36–43; scales 36–53 (usually fewer than 51); longest dorsal spine most frequently 4th or 6th (dorsal-spine filaments never fused with spines), longest dorsal spine 10–21% SL .... *Baldwinella*

9b. Circum-caudal-peduncular scales 34–50; gillrakers 31–39; scales 48–71 (usually 51–67); longest dorsal spine usually 3rd, but 4th or 5th may be longest in individuals less than 120 mm SL, filament of 3rd dorsal spine fused with spine in some larger individuals, longest dorsal spine (or longest dorsal spine plus inseparable filament) 20–53% SL in specimens greater than about 200 mm SL .... *Hemanthias*

10a. Scales 62 or 63; fleshy papillae on posterior half of orbital border 20–22; vomer edentate; anterior naris somewhat remote from posterior naris, internarial distance 2.8–3.1 times in snout length .... *Anatolanthias apiomycter*

10b. Scales 31–57; no fleshy papillae on orbital border; vomer with teeth; anterior naris relatively close to posterior naris, internarial distance 3–21 times (usually more than 5 times) in snout length 11

11a. Accessory scales (secondary squamation) present at bases of larger scales on head or on head and body 12

11b. Accessory scales (secondary squamation) absent 14

12a. Vomerine tooth patch with posterior prolongation; pectoral-fin rays 19–21; in one species posterior margin of caudal fin convex with middle of fin angulated in some specimens (middle rays of fin elongated in some larger individuals), in the other species of the genus two rays of lower caudal-fin lobe greatly elongated; accessory scales present at bases of larger scales .... *Holanthias*

12b. Vomerine tooth patch without posterior prolongation; pectoral-fin rays 16–18; caudal fin lunate to forked—with or without filamentous lobes 13

13a. Scales 46–ca. 50; gillrakers 35–39; soft rays in anal fin 8; circum-caudal-peduncular scales ca. 25 or 26; accessory scales present on head and body .... *Meganthias*

13b. Scales 33–38; gillrakers 42 or 43; soft rays in anal fin 7; circum-caudal-peduncular scales 16–18; accessory scales present on head, absent from body .... *Odontanthias hensleyi*

14a. Lateral line usually interrupted ventral to posterior part of soft dorsal fin; anterior and posterior nares fairly well separated, internarial distance 3–6 times in snout length; circum-caudal-peduncular scales 25–28 (rarely 25); first caudal vertebra with parapophyses .... *Choranthias*

14b. Lateral line continuous, rarely interrupted; anterior and posterior nares closer together, internarial distance 5–21 times in snout length; circum-caudal-peduncular scales 16–28 (usually 17–27); first caudal vertebra without parapophyses 15

15a. Posterior border of anterior naris produced into slender filament; vomerine tooth patch usually with well-developed posterior prolongation, tooth patch may be diamond shaped; tongue usually with teeth .... *Pronotogrammus*

15b. Posterior border of anterior naris produced into short flap but never into long filament; vomerine tooth patch without well-developed posterior prolongation; tongue with or without teeth (present in almost all *A. menezesi*, very frequently in *A. cyprinoides* and *A. nicholsi*, occasionally in *A. asperilinguis*, and rarely in *A. anthias*) .... *Anthias*

ACANTHISTIUS GILL, 1862

Table 1

Acanthistius Gill, 1862:236 (type species *Plectropoma serratum* Cuvier, 1828:399, by monotypy).

Key to Atlantic and Eastern Pacific Species of *Acanthistius*

Accounts of the genus and species are not provided.

1a. Eastern Pacific 2

1b. Atlantic 3

2a. Spines in dorsal fin 11; no dark stripe from tip of snout to orbit and from posterior border of orbit to posterior border of opercle; dark stripe extending from eye to angle of preopercle; area between two upper opercular spines blackish .... *Acanthistius pictus* (Tschudi, 1846) (eastern South Pacific: Santa Rosa, Ecuador, to Valparaiso, Chile [Béarez and Jiménez Prado, 2003])

2b. Spines in dorsal fin 13; dark stripe from tip of snout to orbit and from posterior border of orbit to posterior border of opercle; no dark stripe between eye and angle of preopercle; area between two upper opercular spines not blackish .... *Acanthistius fuscus* Regan, 1913a (eastern South Pacific: Easter and Sala y Gómez islands [Rojas and Pequeño, 1998a])

3a. Spines in dorsal fin 11–13, usually 12, rarely 11 or 13; soft rays in anal fin 7 or 8, usually 7; pectoral-fin rays 18–21, most frequently 20, rarely 18; total gillrakers on first arch, including rudiments, 16–21, most frequently 18 or 19; "head and body buff, covered with small orange spots and few scattered dark brown blotches" (Heemstra and Randall, 1986:510) ....
.... *Acanthistius sebastoides* (Castelnau, 1861) ("Endemic to southern Africa from False Bay, Western Cape Province to Durban. Also reported from Namibia. . . ." [Heemstra, 2010:67], eastern South Atlantic)

3b. Spines in dorsal fin 12 or 13; soft rays in anal fin 8–10; pectoral-fin rays 15–18; total gillrakers on first arch, including rudiments, 20–23 4

4a. Anal-fin rays III, 8; lateral-line scales 56–61; color: light or dark brown with 5 silvery-gray to dark brown vertical bars on flank and caudal peduncle, belly light—whitish to yellowish, body without vermiculated pattern, fins with color pattern similar to that of body (Irigoyen et al., 2008) .... *Acanthistius brasilianus* (Cuvier, 1828) (western South Atlantic: off Brazil from 15°S to 23°S [Irigoyen et al., 2008:52])

4b. Anal-fin rays II or III, 8–10; lateral-line scales 67–70 (Irigoyen et al., 2008); color: dark red, brownish, or gray with small irregularly shaped dark spots on body and on dorsal and anal fins, on body spots usually forming dark bands and irregular vermiculated lines (Nakamura, 1986; Irigoyen et al., 2008) .... *Acanthistius patachonicus* (Jenyns, 1840) (western South Atlantic: from 23°S to 48°S [Irigoyen et al., 2008:52]; rare north of Rio Grande do Sul, Brazil [Carvalho-Filho, 1999])

TRACHYPOMA GÜNTHER, 1859

Table 1

Trachypoma Günther, 1859:167 (type species *Trachypoma macracanthus* Günther, 1859, by monotypy).

Diagnosis. *Trachypoma* is distinguishable from all other genera of Anthiinae treated herein by the following combination of characters. Dorsal-fin rays XII, 13. Anal-fin rays III, 6. Scales mainly cycloid, but some specimens with a number of ctenoid scales—mainly posteriorly (individuals less than about 50 mm SL may have only ctenoid scales). Spines in fins robust. Fins without excessively elongated spines or soft rays.

Description. Mouth terminal or upper jaw exceeding lower jaw very slightly. Premaxillae protrusile. Maxilla falling short of to reaching slightly beyond a vertical through posterior border of orbit. Supramaxilla usually present and well developed. Nares closely set (and near eye) on each side of snout; posterior border of anterior naris with large fimbriate flap that reaches anterior border of posterior naris (or somewhat farther) when reflected. No fleshy papillae on border of orbit. Interorbital region more or less flattened, but supraorbital borders may be somewhat elevated. Scale bone (tabular or extrascapular) with one to three serrae or spinous processes. Dorsal profile anterior to dorsal fin descending rather steeply (up to about a 45-degree angle) to posterior cranial region. Posterior margin of preopercle serrate; ventral margin with two or three (usually three) strong antrorse spines. Opercle with three well-developed spines; dorsalmost spine unusually well developed for an anthiine. Distal margins of interopercle and subopercle without serrae.

Jaws, vomer, and palatines with villiform teeth; vomerine tooth patch chevron shaped, without a posterior prolongation; palatine teeth in a longitudinal band; no teeth on endopterygoids or tongue.

Single dorsal fin, not deeply notched at junction of spinous and soft-rayed parts. Anal fin more or less rounded posteriorly. Pectoral fin rounded posteriorly, roughly symmetrical. Caudal fin rounded; principal rays 17 (9 + 8); branched rays 15 (8 + 7). Vertebrae 26 (10 + 16). First caudal vertebra without parapophyses. Formula for configuration of supraneural bones, etc. usually 0/0+0/2/1+1/1/ (Fig. 2A), rarely 0/0+0/2+1/1/1/. Pleural ribs on vertebrae 3 through 10. No trisegmental pterygiophores associated with dorsal and anal fins.

Scales mostly cycloid (see section on **Squamation** in account of *Trachypoma macracanthus*); no secondary squamation. Interorbital region with embedded scales; dorsum and lateral aspect of snout, maxilla, lower jaw, gular region, branchiostegals, and branchiostegal membranes without scales. Spinous dorsal (at least posteriorly), soft dorsal, anal, and caudal fins with scales basally; pectoral fins with scales on about proximal one-half of their lengths; pelvic fins without scales. Interpelvic process (modified scales overlapping pelvic-fin bases along midventral line) absent. No axillary process associated with pelvic-fin base. Lateral line complete, extending to base of caudal fin (almost straight except gently curving ventrally beneath soft dorsal fin to continue posteriorly near midline of caudal peduncle).

Remarks. Kuiter (2004:125), without explanation, relegated two species of *Acanthistius* (*Acanthistius* sp. and *A. sebastoides*) to *Trachypoma*. We do not agree with those assignments and consider the genus *Trachypoma* to be monotypic.

Trachypoma macracanthus Günther, 1859

Toadstool Groper, Strawberry Cod

Fig. 3; Tables 2–7; Map 1

Trachypoma macracanthus Günther, 1859:167 (original description; four syntypes— BMNH 1855.9.19.40–41, 124 and 144 mm SL, and BMNH 1855.9.19.208–209, 103 and 130 mm SL; type locality off Norfolk Island).

Diagnosis. As for the genus.

Description. Pectoral-fin rays 17 or 18 (usually 17); dorsalmost ray unbranched. Gillrakers (including rudiments) 6 to 8 + 14 to 16—total 20 to 23 (3 to 6 rudiments on upper limb, 3 to 5 on lower limb, total on both limbs 6 to 10). Lateral-line scales 46 to 56. Circum-caudal-peduncular scales ca. 35 to ca. 41 (very difficult to count). Procurrent caudal-fin rays 6 to 8 dorsally, 5 or 6 ventrally. Epineurals associated with first 8 or 9 vertebrae.

Internarial distance usually 8 to 13 times in snout length. Head length 43 to 49% SL. Snout length 8 to 12% SL. Orbit diameter 11 to 13% SL. Body depth at first dorsal spine 36 to 39% SL. Longest dorsal spine (fourth, fifth, or sixth) 18 to 22% SL. Depressed anal fin length 29 to 34% SL. Pelvic-fin length 22 to 27% SL. Upper caudal-fin lobe 22 to 29% SL. Lower caudal-fin lobe 24 to 29% SL.

Squamation. Boulenger (1895) stated that the scales of *Trachypoma macracanthus* are cycloid, but Roberts (1993) reported them to be transforming ctenoid. We found them to be mostly cycloid, but with some ctenoid scales present on most specimens examined. The smallest specimen seen (BPBM 14773, 37 mm SL) has only ctenoid scales. The other specimens examined are much larger (103–168 mm SL); four of these have only cycloid scales; seven have at least some ctenoid scales posteriorly; and the largest (BPBM 36316, 168 mm SL) has many ctenoid scales, mainly below the lateral line anteriorly and on the caudal peduncle posteriorly. Ctenoid scales when present resemble those of serranine serranids (i.e., with rows of ctenial bases present proximal to marginal cteni; see Hughes, 1981) (Fig. 1A). Many scales with smooth posterior borders (as in typical cycloid scales) have ctenial bases in their posterior fields. The preceding suggests that there may be an ontogenetic transformation from ctenoid to cycloid squamation in this species.

Coloration. Günther (1859:168) noted: "The coloration appears now to be brownish olive, being covered all over with round whitish (in life probably blue), dark-edged specks of the size of a scale." Boulenger (1895, plate II) published an excellent black-and-white drawing of *Trachypoma macracanthus*.

A color photograph of *T. macracanthus*, taken underwater, that appears in Francis (1988, plate 33) shows an orange fish with orange narial flaps, red-orange iris, and numerous rather evenly distributed small white spots on head, body, and fins. Kuiter (1993:127) published color photographs, taken underwater, of both juvenile and adult of this species. The adult is similar to the individual depicted by Francis (1988) except that the ground color is yellow orange and the white spotting on the body is less evenly distributed. The juvenile is also similar to the specimen illustrated by Francis (1988) except the body is a mottled dull red and gray.

John E. Randall gave us two color transparencies of specimens of *T. macracanthus*. The following description is based on one of those (BPBM 6613, 156 mm SL, from Easter Island). Ground color of head, body, and fins dull orange overlain with gray ventral to spinous dorsal fin, on basal half of pectoral fin, and in three faint bars—first ventral to anterior part of spinous dorsal fin, second ventral to posterior part of soft dorsal fin, third on caudal peduncle. Head, body, and fins heavily sprinkled with small white spots. Snout almost black. Flap on anterior naris orange. Iris mostly dull orange.

The other transparency received from Randall is a photograph taken underwater (of a specimen observed at Easter Island) that has appeared in print (Randall, 1998:260, fig. 151). It shows a specimen similar in coloration to that described above for BPBM 6613, except there is much less gray on body, flap on anterior naris is dull grayish brown, and many of the spots on head and fins are pale blue.

Maximum length. *Trachypoma macracanthus* reaches a maximum total length of about 400 mm (Francis, 1988; Kuiter, 1993).

Ecological notes. According to Francis (1988:27), *Trachypoma macracanthus* is "most common around offshore islands and coastal headlands" and dwells "under boulders, or in caves and crevices." Kuiter (1993:127) noted that this species occurs in "rocky estuaries to clear coastal waters in shallow protected waters to about 20 m. Usually in boulder reefs, perched in similar fashion to scorpaenid fishes, or in caves." Francis (1988:27) mentioned that "Toadstool groupers rest during the day, and can be approached closely. At night they roam in search of crustaceans, small fishes and shellfish."

The flaps on the posterior borders of the anterior nares are prominently displayed in erect positions in three of the specimens depicted in underwater photographs (Francis, 1988, plate 33; Kuiter, 1993:127). Perhaps these orange structures, moving about with the currents, act as lures for small prey when these fish lie at rest.

Distribution. Yáñez-Arancibia (1975), Randall and Cea Egaña (1984), and Randall et al. (2005) reported *Trachypoma macracanthus* to be a member of the ichthyofauna of Easter Island. Francis (1988:27) wrote that it occurs at the "Kermadec Islands; North Cape to East Cape [North Island, New Zealand]. Abundant at the Kermadecs, rare in New Zealand." Kuiter (1993) gave the distribution as including New South Wales, Lord Howe Island, Norfolk Island, northern New Zealand, and the Kermadec Islands. Pequeño and Lamilla (1996a, b; 2000) and Rojas and Pequeño (1998a) noted that *T. macracanthus* is known from Australia, Lord Howe Island, Norfolk Island, Kermadec Islands, New Zealand, and Easter Island and published the first accounts of it from the Desventuradas Islands off Chile. The specimens that we examined were collected in depths of 1 to 25 meters.

Material examined. Twelve specimens, 37 to 168 mm SL. **EASTER ISLAND:** BPBM 6612 (1 specimen: 166 mm SL), BPBM 6613 (1: 156), BPBM 36316 (1: 168). **NORFOLK ISLAND:** BMNH 1855.9.19.40–41 (2 syntypes: 124–144), BMNH 1855.9.19.208–209 (2 syntypes: 103–130). **LORD HOWE ISLAND:** BPBM 14773 (4: 37–160), SU 9076 (1: 140).

LEPIDOPERCA REGAN, 1914

Table 1

Lepidoperca Regan, 1914:15 (type species *Lepidoperca inornata* Regan, 1914:15, by subsequent designation of Roberts, 1989:561).

Diagnosis. *Lepidoperca* is distinguishable from all other genera of Anthiinae treated herein by the following combination of characters. Scales ctenoid with ctenial bases present proximal to marginal cteni (Fig. 1A). Maxilla scaly. Ventral border of preopercle without antrorse spines. Dorsal-fin rays X, 15 to 21. Lateral-line scales 38 to 51. Total number of gillrakers on first arch 29 to 41.

Description. Mouth more or less terminal, lower jaw usually exceeding upper jaw slightly when mouth closed. Premaxillae protrusile. Maxilla reaching vertical through anterior border of pupil to as far posterior as vertical through posterior border of pupil. Supramaxilla rudimentary or absent. Nares closely set (and near eye) on each side of snout; posterior border of anterior naris produced into a flap but not into a filament. Posterior margin of preopercle finely serrate, ventral margin without antrorse spines.

Fins without extremely elongated spines or soft rays. Single dorsal fin, not deeply notched at junction of spinous and soft-rayed parts; dorsal-fin rays X, 15 to 21. Anal fin usually rounded posteriorly; anal-fin rays III, 7 to 9. Pectoral fin rounded to angulated posteriorly, symmetrical or nearly so, middle rays longest; pectoral-fin rays 14 to 17. Caudal fin lunate to truncate; principal rays 17 (9 + 8); branched rays 15 (8 + 7). Total number of gillrakers on first arch 29 to 41. Lateral-line scales 38 to 51. Vertebrae 26 (10 + 16). Formula for configuration of supraneural bones, etc. 0/0+0/2/1+1/1/ (Fig. 2A). Pleural ribs on vertebrae 3 through 10.

Scales ctenoid (with rows of ctenial bases present proximal to marginal cteni, see Fig. 1A). Head, including snout, maxilla, interorbital, lower jaw, and gular region closely covered with scales. Spinous dorsal fin naked, with scales along margins of spines, or with scales on basal one-third to one-half of fin; other fins scaly basally. Lateral line complete, extending to base of caudal fin, following dorsal body contour below dorsal fin, curving ventrally to continue posteriorly near midlateral axis of caudal peduncle. We have relied heavily on Katayama and Fujii (1982), Roberts (1989), and Andrew et al. (1995) in preparing the above description.

Sexuality, dimorphism, and dichromatism. At least two of the species of *Lepidoperca* are sexually dimorphic in overall size and sexually dichromatic. Roberts (1989:561) wrote that "protogynous hermaphroditism is indicated" for those two species, but he added that "this mode of reproduction has yet to be confirmed by histological analysis of gonad tissue or experimental investigation as described by Sadovy and Shapiro (1987)." On the basis of gross morphology and gonadal histology of another species (*Lepidoperca aurantia*), Roberts (1989:584) noted that "it is reasonable to conclude that... [it] is not sexually dimorphic and is probably a gonochorist."

Remarks. Roberts (1989) revised a group of orange perches from New Zealand and Australian waters that had been confused under the name *Lepidoperca pulchella*. In that paper Roberts provided descriptions for six species, including a new one, and a key to all ten of the species he considered representatives of the genus *Lepidoperca*. Nine of those ten species are native to the New Zealand and Australian region, with only *L. coatsii* from the South Atlantic and the southern Indian Ocean occurring elsewhere. Species of this genus are residents of temperate and subtropical waters near rocky reefs in depths of 10 to 500 meters (Roberts, 1989).

Lepidoperca coatsii (Regan, 1913)

Westwind Drift Orange Perch

Fig. 4; Tables 2–7; Map 2

Caesioperca coatsii Regan, 1913b:237; pl. 6, fig. 1 (original description, illustration; 17 syntypes [four, BMNH 1912.7.1.72–76, 70–78 mm SL; 13 in three lots, NMSZ 1921.143.374—*fide* Eschmeyer, 1998:389]; type locality off Gough Island, South Atlantic Ocean—40°20′S, 09°56′W, at a depth of 183 meters).

Diagnosis. A species of *Lepidoperca* distinguishable from the other species of the genus by the following combination of characters. A dark blotch on membrane posterior to each dorsal spine. Lateral-line scales 42 to 51. Scales in transverse series between lateral line and origin of anal fin 13 (Roberts, 1989:586), 12 to 15 in specimens we examined.

Description. Dorsal-fin rays X, 16 to 19 (most frequently 17). Anal-fin rays III, 7 to 9 (usually 8). Pectoral-fin rays 16 or 17. Gillrakers 9 to 12 + 22 to 25—total 31 to 36. Lateral-line scales 42 to 51 (usually 43–48). Circum-caudal-peduncular scales 17 to 19. Procurrent caudal-fin rays 8 or 9 dorsally, 7 to 9 ventrally. Epineurals associated with first 11 to 13 vertebrae. First caudal vertebra without parapophyses. No trisegmental pterygiophores associated with dorsal and anal fins.

Premaxilla with outer row of conical teeth and inner band of villiform teeth, and with 1 or 2 slightly exserted canines anteriorly; on either side of premaxillary symphysis a patch of villiform to conical teeth, posterior part of patch with a few recurved canines. Dentary with row of conical teeth, near midjaw 1 to 3 recurved canines; anterior to recurved teeth a patch of small conical to villiform teeth adjacent to symphysis; at rear of this patch, a number of posteriorly directed conical to canine teeth; anterior end of jaw with one or two exserted canines. Vomer and palatines with teeth; vomerine tooth patch chevron shaped, without posterior prolongation; palatine teeth in longitudinal band. Endopterygoids and tongue without teeth.

Head length 33 to 40% SL. Snout length 7 to 10% SL. Orbit diameter 11 to 16% SL. Body depth at first dorsal spine 33 to 37% SL. Longest dorsal spine (third or fourth, usually fourth) 17 to 21% SL. Depressed anal-fin length 26 to 31% SL. Pelvic-fin length 23 to 28% SL. Caudal fin slightly emarginate to truncate. Upper caudal-fin lobe 22 to 27% SL. Lower caudal-fin lobe 20 to 25% SL. Andrew et al. (1995) presented a diagnosis for *L. coatsii*.

Coloration. Andrew et al. (1995:pl. 1, fig. H) furnished a color photograph of *Lepidoperca coatsii* and the following notes on its coloration (p. 19):

> Head and body pinkish orange fading to white ventrally; 4 or 5 distinct purple lines running longitudinally along the body below pronounced lateral line. Fins yellowish, except spinous part of dorsal which has dark blotches on interspinous membranes.

Syntypes. The catalogue number (BMNH 1912.7.1.72–76) for four of the syntypes of *Caesioperca coatsii* would lead one to believe that there are (or were) five syntypes in that lot, but PCH recorded data on only four when he had them on loan. Clive Roberts informed (*in litt.* to WDA, 07 February 2001) that only four were present when he examined the lot in 1984. Anthony C. Gill found only four specimens in the jar in the Natural History Museum in London (*in litt.* to WDA, 06 February 2001) and wrote that "either one specimen has gone missing, or someone screwed up when they created the number," and that "there's nothing in the register that might help resolve this."

Eschmeyer (1998:389) noted that there are 13 other syntypes (in three lots) in the collections of the National Museums of Scotland (NMSZ 1921.143.374) (also see Herman et al., 1990:2). We have not seen those specimens.

Distribution. We have examined specimens of *Lepidoperca coatsii* collected off Tristan da Cunha, Nightingale, and Gough islands and from the Austral Seamount. It has been reported from the South Atlantic off the Tristan da Cunha Group and Gough Island in depths of 50 to 183 meters (Regan, 1913b; Penrith, 1967; Andrew et al., 1995), from the Austral Seamount in the southwestern Indian Ocean in 171 to 180 meters (Duhamel, 1984), and from off the southern Indian Ocean islands of Amsterdam and Saint Paul in 50 to 375 meters (Duhamel, 1989, 1997). This species, then, is appropriately considered as a member of the fauna of the West Wind Drift Islands Province, defined by Collette and Parin (1991) as consisting of the Tristan da Cunha Group, Gough Island, Vema Seamount (eastern South Atlantic Ocean), Walters Shoals (southwestern Indian Ocean), Austral Seamount, Amsterdam Island, and Saint Paul Island. Andrew et al. (1995:33), declaring that Collette and Parin (1991) had misinterpreted "the zoogeographic affinities of Walters Shoal and Vema Seamount," excluded those two seamounts from the West Wind Drift Islands Province.

Material examined. Seventeen specimens, 52 to 138 mm SL. **TRISTAN DA CUNHA ISLAND:** USNM 394682 (5 specimens: 52–117 mm SL). **NIGHTINGALE ISLAND, TRISTAN DA CUNHA GROUP:** SAIAB 31523 (3: 110–113), SAIAB 33606 (1: 111). **GOUGH ISLAND:** BMNH 1912.7.1.72–76 (4 syntypes: 70–78), SAIAB 13294 (1: 83), SAIAB 33609 (1: 138). **AUSTRAL SEAMOUNT, SOUTHWESTERN INDIAN OCEAN:** MNHN 1984–82 (1: 90), MNHN 1984–83 (1: 86).

CAPRODON TEMMINCK AND SCHLEGEL, 1843

Table 1

Caprodon Temminck and Schlegel, 1843:64 (type species *Anthias schlegelii* Günther, 1859:93, by subsequent monotypy; type fixed by Günther, 1859:93—genus originally erected without a species; Günther, 1859:88, considered *Caprodon* to be a synonym of *Anthias*).

Neoanthias Castelnau, 1879:366 (type species *Neoanthias guntheri* Castelnau, 1879:367, by monotypy).

Diagnosis. *Caprodon* is distinguishable from all other genera of Anthiinae treated herein by the following combination of characters. Scales ctenoid with ctenial bases present proximal to marginal cteni (Fig. 1A). Maxilla with scales. Ventral border of preopercle without antrorse spines. Dorsal-fin rays X(XI), 19 to 21. Lateral-line scales 55 to 71. Total number of gillrakers on first arch 29 to 41.

Description. When mouth closed, lower jaw exceeding upper jaw. Premaxillae protrusile. Maxilla reaching vertical through middle of orbit or beyond. Supramaxilla absent. Nares closely set (and near eye) on each side of snout; posterior border of anterior naris produced into a flap. Preopercle finely serrate on both limbs, ventral limb without antrorse spines, no spine at angle. Teeth in jaws mostly small, villiform to conical; usually one to a few canines present on each side of symphyses of premaxillae and dentaries.

Fins without extremely elongated spines or soft rays (15th–17th dorsal soft rays moderately elongated in *C. krasyukovae*). Single dorsal fin, not divided to base at junction of spinous and soft-rayed parts. Anal fin usually rounded posteriorly; anal-fin rays III, 7 to 9. Pectoral fin angulated posteriorly with 16 to 19 rays. Caudal fin of variable shape—truncate, slightly emarginate, lunate, forked, or with deep notches dorsal and ventral to elongated median rays; principal caudal-fin rays 17 (9 + 8); branched rays 15 (8 + 7). Vertebrae 26 (10 + 16), except 27 in the only described specimen of *C. krasyukovae*.

Scales ctenoid (with rows of ctenial bases present proximal to marginal cteni; see Hughes, 1981) (Fig. 1A); no secondary squamation. Most of head, including maxilla, dorsal and lateral aspects of snout, lachrymal region, interorbital region, and most of lower jaw heavily covered with scales. Each fin heavily covered with scales for considerable distance distal to its base. Lateral line complete, extending to at least base of caudal fin, running parallel to dorsal body contour below dorsal fin, curving to near midlateral axis of body on caudal peduncle.

Species of *Caprodon*. Five nominal species have been assigned to the genus *Caprodon*: *C. schlegelii* (Günther, 1859*), C. longimanus* (Günther, 1859), *C. affinis* Tanaka, 1924, *C. unicolor* Katayama, 1975, and *C. krasyukovae* Kharin, 1983a. In view of the confusion discussed below, how many of those nominal species are valid is uncertain.

Günther (1859:93–95) described *Caprodon schlegelii*, from the "Japanese Sea," and *C. longimanus*, locality unknown, in the same publication, but did not clearly distinguish the two. In the original description of *C. affinis*, based on a specimen obtained from the market in Tokyo, Tanaka (1924:612) wrote that his new species was "Essentially similar to the preceding species" (i.e., *C. schlegelii*) and even entertained the idea (p. 613) that it might be the female of that species, but he could "not consider that the differences are due to sex or age or color variation." Later Tanaka (1931), Katayama (1960), and Kharin and Dudarev (1983) considered *C. affinis* to be a junior synonym of *C. schlegelii*. Katayama (1960:137) concluded "with some doubt" that *C. longimanus* "is no other than female" and that *C. schlegelii* "probably male" of the same species, *C. schlegelii*.

In 1975 Katayama described *C. unicolor* from two specimens collected off Midway Island and stated (p. 13) that "The present new species is closely related to *Caprodon schlegelii*. . . ." Katayama mentioned five differences between the two species, the most significant being numbers of gillrakers on the first arch—total on that arch 35 to 37 in *unicolor*, 29 to 32 in *schlegelii*. (Randall, 2007, gave total gillrakers as 33 to 35 in *unicolor* and 32 to 34 in *schlegelii*. Randall also confirmed another distinguishing character cited by Katayama, finding head length to be 3.1–3.25 in SL in *unicolor* vs. 2.8–3.1 in SL in *schlegelii*.) Kharin and Dudarev (1983:20, 24) placed *C. unicolor* in the synonymy of *C. longimanus* because they could find "no substantial differences between" Katayama's description of *C. unicolor* and *C. longimanus*.

Kharin (1983a) described *C. krasyukovae* from a specimen collected over the Lord Howe Rise, and Kharin and Dudarev (1983:23–24) considered both *C. schlegelii* and *C. longimanus* to be valid species and provided a key to separate the three species of *Caprodon* they recognized. According to those authors (p. 24), *krasyukovae* can be distinguished from the other two species by its possession of unusually long 15th, 16th, and 17th soft rays in the dorsal fin, two deep notches in distal margin of caudal fin—one dorsal to and one ventral to "strongly elongated median rays," and 27 total vertebrae vs. all soft rays of dorsal fin of about the same length, caudal fin of a different shape, and 26 total vertebrae. Kharin and Dudarev (1983:24) separated *C. longimanus* from *C. schlegelii* by caudal-fin shape and gillraker count—*longimanus* having a caudal fin "with a deep median notch" and "9–11 + 24–29 gill rakers on the first gill arch" in contrast with *schlegelii* with a caudal fin "without or with a very weak median notch" and "8–10 + 21–23 gill rakers on first gill arch." Interestingly, Kuiter (2004:100) published six color photographs of specimens ranging in size from 180 to 430 mm that he identified as *C. krasyukovae* and gave the range of that species as "Lord Howe Island region to southern Qld and NSW coasts."

Randall (1995:196) included *Caprodon schlegelii* in a list of fishes "shared by the Hawaiian Islands and the region from southern Japan to Taiwan, but not areas to the south. . . ." In contrast, *C. longimanus* (*sensu lato*) is widely distributed in the South Pacific from Australia to Chile (see sections on **Comparisons of populations of Caprodon longimanus** and **Distribution** under the account for *Caprodon longimanus*). These two species may form an anti-equatorial pair. More recently Randall (2007:190) reassessed the Hawaiian *Caprodon* and concluded that it is conspecific with Katayama's *C. unicolor* from Midway Island.

Caprodon longimanus (Günther, 1859)

Long-finned Perch; Pink Maomao

Fig. 5; Tables 2–7; Map 1

Anthias longimanus Günther, 1859:94 (original description; holotype, BMNH 1857.6.13.116, stuffed specimen of ca. 250 mm SL; type locality unknown).

Caprodon longimanus: Meléndez and Villalba, 1992:9, fig. 13 (description, illustration, Juan Fernández Islands).---Rojas and Pequeño, 1998a:177, fig. 4 (description, illustration; Desventuradas Islands—San Ambrosio and San Félix, Juan Fernández Islands, Easter Island, South Pacific, New South Wales).—Rojas and Pequeño, 1998c:44, fig. 2, table 1 (description, illustration; Desventuradas Islands—San Ambrosio and San Félix, Juan Fernández Islands, Easter Island, South Pacific, New South Wales).

Diagnosis. A species of *Caprodon* distinguishable from the other species of the genus by the following combination of characters. Caudal fin emarginate to forked. All rays in posterior part of soft dorsal fin of about same length, none noticeably elongated. Vertebrae 26. Total gillrakers on first arch 34 to 37 (33–41 in specimens from the western Pacific).

Description. This description is based only on the examination of specimens from the eastern Pacific. Dorsal-fin rays X, 19 to 21 (most frequently 20). Anal-fin rays III, 8. Pectoral-fin rays 16 to 18 (most frequently 17). Gillrakers 8 to 11 + 25 to 28—total 34 to 37. Lateral-line scales 58 to 65 (usually 61–64). Circum-caudal-peduncular scales 24 to 30. Procurrent caudal-fin rays 9 to 11, both dorsally and ventrally. First caudal vertebra without parapophyses. Formula for configuration of supraneural bones, etc. 0/0+0/2/1+1/1/ (Fig. 2A). Pleural ribs on vertebrae 3 through 10. Epineurals associated with first 8 or 9 vertebrae. No dorsal trisegmental pterygiophores; anal trisegmental pterygiophores 0 to 2.

Canines at anterior ends of jaws somewhat exserted on premaxillae, strongly exserted on dentaries; one to a few canines somewhat posterior to anterior end of each dentary. Vomer, palatines, endopterygoids, and tongue with teeth; patch of teeth on vomer with broad posterior prolongation.

Internarial distance 13 to 30 times in snout length. Head length 28 to 31% SL. Snout length 5 to 8% SL. Orbit diameter 8 to 10% SL. Body depth at first dorsal spine 29 to 33% SL. Longest dorsal spine (fifth or sixth) 13 to 17% SL. Depressed anal fin length 25 to 27% SL. Pelvic-fin length 18 to 23% SL. Upper caudal-fin lobe 27 to 35% SL. Lower caudal-fin lobe 24 to 30% SL.

Coloration. In the original description, Günther (1859:95) wrote that "The specimen is now discoloured, but appears to have had a red ground-colour." Boulenger (1895:315) stated that *Caprodon longimanus* is "Pink, with yellow spots or irregular oblique stripes; two yellow stripes in front of the eye, and another running obliquely from below the eye to the base of the pectoral; dorsal and anal yellow and red, the former sometimes with irregular black blotches." Boulenger (1895, plate XII) also provided a fine black and white drawing of *C. longimanus*.

Katayama (1960) furnished two beautiful color illustrations labeled as the male (plate 17) and female (plate 18) of *Caprodon schlegelii*. In view of the fact that Katayama (1960:137) concluded "with some doubt" that some specimens (presumably from Japanese waters) examined by him that were referable to *C. longimanus* are females of *C. schlegelii*, his plate 18 (Katayama, 1960) may be of the species herein considered as *C. longimanus*, but due to the unsettled state of the taxonomy of *Caprodon* this is uncertain.

Burgess and Axelrod (1973:496) presented color illustrations labeled *Caprodon schlegelli* (plate 428) and *C. longimanus* (plate 429). The fish in plate 429 is mainly orange dorsolaterally and pink ventrolaterally with five roughly rectangular greenish blotches dorsolaterally—anteriormost blotch beneath first three spines in dorsal fin, second below last dorsal spine and first eight dorsal soft rays, next two beneath posterior two-thirds of soft dorsal fin, and posteriormost a saddle on caudal peduncle; head orange dorsally and pink ventrally; ventral half of iris of eye yellow, dorsal half gray; spinous dorsal fin mainly yellow green; soft dorsal fin mainly pink with orange border distally and orange base that is partly overlain by three fairly large roughly rectangular yellow green blotches; anal fin mostly pink, with orange border distally and band of yellow near mid-fin extending from second spine to fifth soft ray; pectoral fin pink, with yellow distal border along dorsal half of fin; pelvic fin pink; caudal fin mostly pink, with orange extending distally from base of fin for some distance between the rays.

Francis (1988:26) described the coloration as:

Bright pink with red-orange markings on face, and light blue margins on fins. A rare colour variety (possibly the spawning male) has black markings on rear of dorsal fin and upper back, and lines of yellow spots on face and upper front part of body. All fins yellow except dorsal, and all have blue margins or blue tips except pectoral.

Francis (1988, plates 40 & 41) also presented photographs of the two color morphs.

Kuiter (1993:136) furnished an underwater color photograph of an animal identified as *Caprodon longimanus*, presumably encountered off southeastern Australia. The coloration of the fish in that photograph closely resembles that of the "rare colour variety" depicted in plate 41 of Francis (1988). The individual in Kuiter's photograph has a purple ground color and many flecks of yellow and gold on body and posterior part of head; anterior part of head with yellow and gold vermiculations; ground color of spinous dorsal and anal fins largely purple overlain with many yellow spots; anterior part of soft dorsal fin mostly black; posterior part of soft dorsal and caudal fins mostly yellow green; pectoral fin with grayish purple ground color and yellow flecks; leading edge of pelvic fin blue, remainder mostly yellow; diffuse black bar extending ventrally beneath black part of soft dorsal fin to a point well below lateral line.

Comments on the holotype. The type locality of *Anthias longimanus* is unknown, but perhaps, as suggested by Günther (1859:95), "its native sea may be some part of the Indian or Australian seas." The following counts are taken from the holotype (BMNH 1857.6.13.116, ca. 250 mm SL): dorsal-fin rays X, 20; anal-fin rays III, 8; pectoral-fin rays ca. 16 (both sides); principal caudal-fin rays 17 (9 + 8); lateral-line scales ca. 58 (left), ca. 60 (right)—a few scales missing on each side (the numbers missing were estimated and included in the recorded counts). Being stuffed, the holotype is somewhat distorted, particularly in depth of body. Consequently, some measurements of this specimen are not reliable enough to use for comparative purposes.

Comparisons of populations of *Caprodon longimanus*. We have examined specimens from both the eastern and western Pacific and found differences in a few countable characters, most strikingly in circum-caudal-peduncular scales (24–30, mean 26.6, in eastern Pacific specimens; 30–34, mean 31.9, in western Pacific material). Differences were also observed in lateral-line scales (58–65, mean 62.1, eastern Pacific; 61–71, mean 65.5, western Pacific) and total gillrakers (34–37, mean 35.2, eastern Pacific; 33–41, mean 37.6, western Pacific). Additionally, depth of body at first dorsal spine varies between the populations—29 to 33% SL in eastern Pacific specimens, 32 to 38% SL in western Pacific individuals. Although the sample sizes are relatively small (17 specimens, 108–276 mm SL, from the eastern Pacific; 33, 102–380 mm SL, from the western Pacific), the variations observed are apparently indicative of real differences between the populations and suggest that there are two species currently included under the name *Caprodon longimanus*.

Synonyms. In view of the uncertainty concerning the number of valid species of *Caprodon* (see **Species of Caprodon** under the account of *Caprodon*) and the fact that we have examined the type of only one of the nominal species (viz., *C. longimanus*) of the genus, we have not attempted to prepare a detailed synonymy for *C. longimanus*. It is important to note, however, that Boulenger (1895) placed both *Scorpis fairchildi* Hector, 1875, and *Neoanthias guntheri* Castelnau, 1879, in the synonymy of *C. longimanus* and to reiterate that Kharin and Dudarev (1983) considered *C. unicolor* Katayama, 1975, to be a synonym of *C. longimanus*. Randall (2007), however, considered *C. unicolor* to be a valid species.

Ecological notes. Francis (1988:26) mentioned that in New Zealand waters *C. longimanus* is "most abundant around islands and coastal headlands" and that it "schools in midwater in areas of moderate current flow around islands, pinnacles and archways." He further mentioned that it rests on rocky bottom at night with a color change to "blotchy pink-red" and opined that it is probably a winter spawner.

Francis (1988:26) noted that *C. longimanus* feeds "actively during the day on plankton and salps carried by currents," and occasionally feeds at the surface but is usually found in depths greater than 10 meters. Parin et al. (1997:195) found micronekton to be the predominant type of food consumed by *C. longimanus* in the Nazca/Sala y Gómez region of the eastern Pacific. Rojas et al. (1998) studied the diet and food preference of 55 specimens (176–286 mm SL) of this species collected off Alejandro Selkirk Island (Juan Fernández Islands) and found the pelagic tunicate *Thalia* sp. to be the most frequent and abundant prey; other food items included radiolarians, siphonophores, polychaetes, pteropods, crustaceans (17 genera), and chaetognaths. They (Rojas et al., 1998:939) characterized *C. longimanus* as being an "opportunistic pelagic polyphagic-zooplanktivorous predator." The second author found that the gut of a specimen (QM 13419, 147 mm SL) collected in the Tasman Sea was full of pelagic tunicates.

Distribution. Because the number of valid species in the genus *Caprodon* is unknown, it is impossible to give precise limits to the geographic range of *C. longimanus*. We have examined specimens from the eastern Pacific (Juan Fernández Islands, near the southwest end of the Nazca Ridge, and Easter Island—examined by J. E. Randall) and from the western Pacific (Tasman Sea and New South Wales, Australia) collected in depths of 0 (0/20) to 402 meters. Randall and Cea Egaña (1984) and Randall et al. (2005) included this species in lists of fishes known from Easter Island. Parin (1991) listed *C. longimanus* in a table of fishes recorded from the Nazca and Sala y Gómez ridges, and Parin et al. (1997) recorded it in depths of 160 to 235 meters on Seamount 12 (Bolshaya), a guyot in the transitional Nazca/Sala y Gómez area. Meléndez and Villalba (1992) examined six specimens (191–245 mm SL) collected in depths of 1 to 15 meters in the Juan Fernández Islands. Pequeño and Lamilla (1996a, b) and Pequeño and Sáez (2000) noted that *C. longimanus* is known from the Desventuradas Islands off Chile and from the continental coast of Chile. Rojas and Pequeño (1998a, c) examined specimens collected off Australia, New Zealand, Easter Island, Juan Fernández Islands, and Desventuradas Islands.

Material examined. Fifty-one specimens, 102 to 380 mm SL. **LOCALITY UNKNOWN:** BMNH 1857.6.13.116 (holotype: stuffed, ca. 250 mm SL). **JUAN FERNÁNDEZ ISLANDS:** MCZ 46166 (5 specimens: 108–264), USNM 176579 (2: 211–224), USNM 176580 (1: 177), USNM 176581 (1: 159). **NEAR SOUTHWEST END OF NAZCA RIDGE:** BPBM 36432 (1: 200), BPBM 36440 (3: 193–204), GMBL 79–258 (3: 200–212). **EASTER ISLAND:** BPBM 6644 (1: 276). **TASMAN SEA, EAST OF LORD HOWE ISLAND:** QM 13419 (18: 102–153). **NEW SOUTH WALES:** AMS I. 18423–001 (1: 271), AMS I. 18788–001 (1: 346), AMS I. 21447–003 (12: 150–225), PC Heemstra's collection No. 47 (1: 380)—apparently lost.

HYPOPLECTRODES GILL, 1862

Table 1

Hypoplectrodes Gill, 1862:236 (type species *Plectropoma nigrorubrum* Cuvier, 1828, by monotypy).

Gilbertia Jordan, 1890:346 (type species *Plectropoma semicinctum* Valenciennes, 1833, by original designation; preoccupied by *Gilbertia* Cossman, 1889, a genus of Mollusca).

Ellerkeldia Whitley, 1927:298 (type species *Plectropoma semicinctum* Valenciennes, 1833, by virtue of the facts that *Ellerkeldia* was proposed as a replacement name for *Gilbertia* Jordan, 1890, preoccupied by *Gilbertia* Cossman, 1889, and that a replacement name retains the type of the prior name; but see discussion of this below).

Scopularia de Buen, 1959:95 (type species *Scopularia rubra* de Buen, 1959 [= *Plectropoma semicinctum* Valenciennes, 1833], by original designation).

Diagnosis. *Hypoplectrodes* is distinguishable from all other genera of Anthiinae treated herein by the following combination of characters. Scales ctenoid with ctenial bases present proximal to marginal cteni (Fig. 1A). Preopercle with one to three antrorse spines on ventral border. Maxilla usually without scales. Dorsal-fin rays X(XI), 16 to 22. Lateral-line scales 40 to 65. Total number of gillrakers on first arch 17 to 22. Total vertebrae usually 27 (occasionally 28, rarely 26).

Description. Mouth terminal or lower jaw exceeding upper jaw slightly when mouth closed. Premaxillae protrusile. Maxilla reaching (or almost reaching) vertical through posterior margin of orbit or slightly beyond when mouth closed. Supramaxilla typically present, sometimes quite small. Nares closely set (and near eye) on each side of snout; posterior border of anterior naris produced. Posterior margin of preopercle serrate; one to three antrorse spines on preopercle (one spine usually at angle or on ventral margin near angle, other spine(s) on ventral margin).

Fins without extremely elongated spines or soft rays. Single dorsal fin not divided to base at junction of spinous and soft-rayed portions, but fin may appear notched at junction; dorsal-fin rays X(XI), 16 to 22. Margin of anal fin broadly rounded to squared off posteriorly; anal fin rays III, 7 to 9. Pectoral fin symmetrical, middle rays longest; pectoral-fin rays 13 to 18. Caudal fin truncate to slightly rounded; principal rays 17 (9 + 8); branched rays 15 (8 + 7); procurrent rays 7 to 10 dorsally, 6 to 9 ventrally. Gillrakers not very numerous, total on first arch 17 to 22; both limbs of first arch typically with a number of rudiments. Lateral-line scales 40 to 65. Circum-caudal-peduncular scales 27 to 34. Vertebrae usually 27 (10 + 17), occasionally 28 (10 + 18) in the type species—*H. nigroruber*, rarely 10 + 16 = 26 in *H. maccullochi*. First caudal vertebra without parapophyses. Epineurals associated with first 8 to 10 vertebrae. Pleural ribs on vertebrae 3 through 10, rarely present on eleventh vertebra in *H. nigroruber*. No trisegmental pterygiophores associated with dorsal fin, 0 to 5 associated with anal fin. Formula for configuration of supraneural bones, etc. 0/0+0/2/1+1/1/ (Fig. 2A).

Scales ctenoid (with rows of ctenial bases present proximal to marginal cteni; see Hughes, 1981) (Fig. 1A). Lateral line complete, extending to at least base of caudal fin, running parallel to dorsal body contour below dorsal fin, then curving to near midlateral axis of body on caudal peduncle.

Coloration. Kuiter (2004) published excellent color photographs of all known species of *Hypoplectrodes* (including two undescribed ones).

Nomenclatural considerations. Whitley (1927:298), in presenting his new name *Ellerkeldia*, wrote: "This name is proposed for the Australian fish *Gilbertia annulata* (Günther), originally described as a *Plectropoma* (Cat. Fish. Brit. Mus., i, 1859, p. 158)." Further on he stated that Jordan's *Gilbertia* (type species *Plectropoma semicinctum* Valenciennes, 1833) "is thus clearly preoccupied by *Gilbertia* Cossman . . . a genus of Mollusca, and *Ellerkeldia* should be used in its stead." At the end of his comments on *Ellerkeldia*, Whitley specified *Ellerkeldia annulata* (Günther) as the type species (orthotype) of that genus. In their synonymy for *Hypoplectrodes*, Anderson and Heemstra (1989:1002) wrote: "Whitley, 1927, incorrectly considered *Plectropoma annulatum* Günther, 1859, as the type species" [of *Ellerkeldia*]. That statement was based on the following edict:

> If an author publishes a new scientific name expressly as a replacement for a previously established genus-group name (nomen novum) . . . both the prior nominal taxon and its replacement must have the same type species, and . . . type fixation for either applies also to the other, despite any statement to the contrary (ICZN, 1985:125, Article 67h; see also ICZN, 1999:68, Article 67.8).

Eschmeyer (1990:135; 1998:1926) disagreed with Anderson and Heemstra's (1989) interpretation and considered Whitley's designation of *Plectropoma annulatum* as the type species a valid act. Further on, Eschmeyer (1990:660; 1998:2868–2869) discussed in some detail his rationale for accepting Whitley's type designation. His argument is interesting, but not persuasive in view of Article 67.8 (ICZN, 1999) and Whitley's statement that "*Ellerkeldia* should be used in its [*Gilbertia*'s] stead." We conclude that "should be used in its stead" is reasonably interpreted (in the sense of Article 67.8) as "expressly" creating "a new replacement name (nomen novum) for a previously established name. . . ."

Anderson and Heemstra (1989:1003) noted that:

> Generic names such as *Hypoplectrodes*, with the suffix "-odes," are substantivated adjectives and are masculine (ICZN [1985] Article 30b). Accordingly, adjectival specific names in combination with *Hypoplectrodes* must have the masculine termination (ICZN [1985], Article 31b).

(Also see ICZN, 1999, Articles 30.1.4.4 and 31.2.)

In addition to *H. semicinctum*, covered herein, there are nine other known species (two undescribed) in the genus, all from temperate seas off either Australia or New Zealand. Allen and Moyer (1980:332) presented a key to six of these (as species of *Ellerkeldia*), and Anderson and Heemstra (1989) gave the literature citations for the original descriptions of all described species. Anderson and Heemstra (1989) synonymized *Scopularia rubra* with *Hypoplectrodes semicinctum*. As a result of this action, *Ellerkeldia rubra* Allen, 1976 (= *Hypoplectrodes ruber*), became a junior secondary homonym of *H. ruber* (de Buen, 1959). Allen and Randall (1990) published a replacement name, *H. cardinalis*, for *H. ruber*.

Comments on the taxonomy of *Hypoplectrodes*. Anderson and Heemstra (1989) treated the increased number of vertebrae (27 or 28) in species of *Hypoplectrodes* as a synapomorphy that differentiates this genus from other anthiines which usually have 26 vertebrae.

Randall (1980:102) considered *Ellerkeldia* (= *Hypoplectrodes*) to be "closely related to *Plectranthias*" Bleeker, 1873, but provided no evidence in support of that assertion. Because there are at least 10 species of *Hypoplectrodes* (eight with names, two undescribed) and many species relegated to *Plectranthias* (48 described and others undescribed that at present would be assigned to *Plectranthias* [Randall and Hoese, 1995; Randall, 1996; Zajonz, 2006; Anderson, 2008; Heemstra and Randall, 2009; Wu et al., 2011]), the evaluation of the relationship of *Hypoplectrodes* to *Plectranthias* will require much additional work.

Miscellaneous notes on the biology of *Hypoplectrodes*. Two species of *Hypoplectrodes*, *H. huntii* (collected at Poor Knights Islands off northeastern North Island, New Zealand, and reported as *Ellerkeldia huntii*; Jones, 1980) and *H. maccullochi* (from the coast of New South Wales; Webb and Kingsford, 1992) have been reported to be protogynous hermaphrodites. Baldwin and Neira (1998) provided a description of the larvae of *H. maccullochi* based on seven specimens (5.2–8.4 mm body length) collected off New South Wales. Jones (1980) reported that individuals of *H. huntii* mainly consumed gammarid amphipods and mysid shrimps as they foraged in the algal canopy.

Hypoplectrodes semicinctum (Valenciennes, 1833)

Semigirdled Perch

Fig. 6; Tables 2–7; Map 3

Plectropoma semicinctum Valenciennes, 1833:442 (original description; holotype MNHN 0000–7777, 146 mm SL; type locality Juan Fernández Islands, eastern South Pacific).

Hypoplectrodes semicinctum: Meléndez and Villalba, 1992:10, fig. 14 (description, illustration).—Rojas and Pequeño, 1998a:182, fig. 6 (description, illustration).—Rojas and Pequeño, 1998c:46, fig. 3, table 2 (description, illustration).

Scopularia rubra de Buen, 1959:95, fig. 14 (original description, illustration; holotype and paratype EBMC 123–124, 174 and 131 mm TL, respectively, apparently lost; type locality Cumberland Bay, Más a Tierra Island, Juan Fernández Islands, eastern South Pacific).

Diagnosis. A species of *Hypoplectrodes* distinguishable from the other described species of the genus by the following combination of characters. Lateral-line scales 48 to 55. Dorsal-fin rays X, 19 to 22. Pectoral-fin rays 15 to 18. Gillrakers on lower limb of first arch: developed—8 or 9; rudimentary—4 to 6. Second anal spine very well developed, its length 15 to 20% SL. Body usually with 9 darkly pigmented bars (including one on nape); bars wider than lightly pigmented interspaces.

Description. Dorsal-fin rays X, 19 to 22 (usually 20 or 21). Anal-fin rays III, 7 to 9 (usually 8, rarely 7 or 9). Pectoral-fin rays 15 to 18 (rarely 15 or 18). Gillrakers including rudiments on first arch 4 to 6 + 12 to 14—total 17 to 20 (developed gillrakers on upper limb 1, on lower limb 8 or 9). Lateral-line scales 48 to 55 (usually 49–51). Circum-caudal-peduncular scales 27 to 32 (usually 28–30). Procurrent caudal-fin rays 8 to 10 (very rarely 10) dorsally, 6 to 9 (usually 7 or 8) ventrally. Vertebrae 27 (10 + 17). Epineurals associated with first 9 or 10 vertebrae (usually first 9). Anal trisegmental pterygiophores 0 to 3 (most frequently 1).

Premaxilla with band of small conical teeth; band expanded anteriorly; posterior teeth in expanded part of band (adjacent to symphysis) enlarged and posteriorly directed; one or two canines at anterior end of jaw; no teeth at symphysis. Dentary with band of small conical teeth; band somewhat expanded adjacent to symphysis; one to three canines at about middle of band; numerous enlarged posteriorly directed conical teeth at anterior end of band near symphysis; one or two canines (may be exserted) at anterior end of jaw; no teeth at symphysis. Vomer and palatines with small conical teeth; vomerine tooth patch chevron shaped, without a posterior prolongation; palatine teeth in a longitudinal band; no teeth on endopterygoids or tongue.

Ventral margin of preopercle with two or three, usually three, antrorse spines; spines frequently covered by skin. Most of head covered with scales; dorsum and lateral aspect of snout, maxilla, supramaxilla, lower jaw, branchiostegal membranes, and most of branchiostegals without scales; gular region usually without scales; squamation variously developed on interopercle, but usually confined to posterior part. Squamation well developed on bases of all fins and continuing for some distance onto fins. No axillary process at base of pelvic fin. No smaller accessory scales (squamulae) at bases of body scales.

Head length 39 to 45% SL. Snout length 10 to 13% SL. Orbit diameter 6 to 10% SL. Body depth at first dorsal spine 33 to 37% SL. Longest dorsal spine (fourth, fifth, or sixth—usually fifth) 14 to 18% SL. Depressed anal fin length 29 to 32% SL. Pelvic-fin length 21 to 26% SL. Upper caudal-fin lobe 21 to 25% SL. Lower caudal-fin lobe 21 to 26% SL. Anderson and Heemstra (1989) presented a detailed description of *H. semicinctum.*

Coloration. For a detailed description of the coloration in alcohol, see Anderson and Heemstra (1989:1013). Valenciennes (1833:443) described the coloration of the holotype of *Plectropoma semicinctum.* He wrote:

> *Les couleurs de ce poisson sont un beau rouge vermillon, traversé par huit demi-bandes d'un rouge-brun vif, qui descendent du dos et s'arrêtent sur le milieu des côtés, de manière à former des demi-ceintures sur les flancs. La dernière seule entoure presque la queue entière. Des traits bruns plus pâles et obliques traversent les joues, et forment sur l'opercule des rivulations confuses. La dorsale et la caudale sont rougeâtres. La pectorale, les ventrales et l'anale ont de l'olivâtre, mêlé dans le rouge qui fait le fond général de la couleur.*

The colors of this fish are a beautiful vermilion red, traversed by eight half bands of a bright red brown that descend on the back and stop on the middle of the sides, so as to form half belts on the flanks. Only the last almost encircles the entire tail. Some paler and oblique brown bars cross the cheeks and form on the opercle indistinct rivulations. The dorsal and caudal are reddish. The pectoral, ventrals, and anal are olive, mixed with the red that forms the general background color. (Translated by WDA; Anderson and Heemstra, 1989:1013.)

De Buen (1959) stated that *Scopularia rubra* (= *H. semicinctum*) is red with black bands. Kuiter (2004) presented a fine color photograph of *H. semicinctum*.

Orthography. Anderson and Heemstra (1989:1014) discussed the correct spelling of the specific name and considered it to be *semicinctum*.

Distribution. We have examined specimens of *H. semicinctum* collected in the eastern South Pacific off the Juan Fernández Islands and San Félix Island (one of the Desventuradas Islands) in shallow waters with a maximum depth of 20 meters. Meléndez and Villalba (1992) reported six specimens (116–175 mm SL) of *H. semicinctum* collected in depths of 5 to 15 meters in the Juan Fernández Islands. Pequeño and Lamilla (1996b; 2000), Rojas and Pequeño (1998a, c), and Pequeño and Sáez (2000) also reported this species from Juan Fernández and from the Desventuradas.

Kuiter (2004:116) gave the distribution of *H. semicinctum* as "Juan Fernández Islands, where common, rare in the Galápagos Islands and a single record from Easter Island." Concerning Kuiter's report of *H. semicinctum* occurring in the Galápagos, McCosker and Rosenblatt (2010:191) wrote: "We are unaware of any specimen from Galápagos or any other published record" from the Galápagos.

Yáñez-Arancibia (1975:38, fig. 2) illustrated a specimen labeled *Scopularia rubra* and considered to be from Easter Island. This illustration is a good likeness of *H. semicinctum*. Based on Yáñez-Arancibia's (1975) report of *Scopularia rubra*, Randall and Cea Egaña (1984) included *Ellerkeldia rubra* (de Buen) (= *H. semicinctum*) in their paper on native names of Easter Island fishes. Randall (*in litt.* to WDA, ca. 1988) informed that he has neither observed nor collected *H. semicinctum* at Easter Island, despite having collected fishes extensively there on three separate occasions; in addition, he has not met any fishermen or divers there who are familiar with this species. As a consequence, Randall does not think that there is a breeding population of *H. semicinctum* at Easter Island (at least not in shallow water) and that Yáñez-Arancibia's record was probably of a stray or possibly of a specimen for which the locality was recorded incorrectly (Anderson and Heemstra, 1989). Despite the preceding, Rojas and Pequeño (1998a, c) listed a 175-mm SL specimen (IZUC 2991) of *H. semicinctum* from Easter Island in their material examined. This specimen was also mentioned by Pequeño and Lamilla (1996b:28), who noted "La etiqueta indica: 'presumido Isla de Pascua'" (= "The label states: 'presumed to be from Easter Island'"). Whether this specimen was a stray that wandered to Easter Island or was even collected in that locality is open to conjecture. Randall et al. (2005) put *H. semicinctum* in a checklist of fishes recorded from Easter Island, and later Randall (*in litt.* to WDA, 25 August 2008) wrote that he did not "recall ever seeing a specimen of *H. semicinctum* from anywhere," but "even if just a waif" at Easter Island it would have been included in the checklist. More recently Randall and Cea (2011:57) noted that "There is one record of . . . *Hypoplectrodes semicinctum* . . . from Easter Island, but it appears to be a locality error."

Material examined. In addition to the 38 specimens (38–177 mm SL) listed by Anderson and Heemstra (1989), we have examined an additional specimen from San Félix Island: IZUC 2768 (151 mm SL).

PLECTRANTHIAS BLEEKER, 1873

Table 1

Plectranthias Bleeker, 1873:238 (type species *Plectropoma anthioides* Günther, 1872:655, by monotypy).

Paracirrhites Steindachner, 1883:25 (type species *Paracirrhites japonicus* Steindachner, 1883:25, by monotypy; objectively invalid, being preoccupied by *Paracirrhites* Bleeker, 1874, in fishes; replaced by *Isobuna* Jordan, 1907).

Sayonara Jordan and Seale, 1906:145 (type species *Sayonara satsumae* Jordan and Seale, 1906:145, by original designation and monotypy).

Isobuna Jordan, 1907:158 (type species *Paracirrhites japonicus* Steindachner, 1883:25, by *Isobuna* being a replacement name for *Paracirrhites* Steindachner, 1883).

Xenanthias Regan, 1908:223 (type species *Xenanthias gardineri* Regan, 1908:223, by monotypy).

Zalanthias Jordan and Richardson, 1910:470 (proposed as a subgenus; type species *Anthias kelloggi* Jordan and Evermann, 1903:179, by original designation; Smith and Craig, 2007:45, included *Zalanthias kelloggi* in Serranidae: Serraninae).

Pteranthias Weber, 1913:208 (type species *Pteranthias longimanus* Weber, 1913:209, by monotypy).

Serranops Regan, 1914:15 (type species *Serranops maculicauda* Regan, 1914:15, by monotypy).

Pelontrus Smith, 1961:364 (type species *Pelontrus morgansi*, Smith, 1961:365, by original designation and monotypy).

Zacallanthias Katayama, 1964:27 (type species *Zacallanthias sagamiensis* Katayama, 1964:28, by original designation and monotypy).

Diagnosis. The species of *Plectranthias* included herein are distinguishable from the species of all other genera of Anthiinae covered in this work by the following combination of characters. Dorsal-fin rays X, 15 to 17. Anal-fin rays III, 7. Pectoral-fin rays 12 to 17. Total gillrakers on first arch 14 to 31. Lateral line complete, not interrupted, tubed scales 28 to 46. Circum-caudal-peduncular scales 12 to 22. Vomerine tooth patch chevron shaped without posterior prolongation. No secondary squamation. Maxilla usually without scales.

Description. The following applies to the species of *Plectranthias* included herein. Premaxillae protrusile. Small splint-like supramaxilla usually present. Anterior and posterior nares rather closely set on each side of snout; anterior naris at distal end of short tube, posterior border of anterior naris produced into a short flap but never into a long filament; internarial distance 7 to 17 times in snout length. No fleshy papillae on border of orbit. Preopercle serrate; lower limb with or without antrorse spines.

Teeth in jaws mostly conical, anteriorly some enlarged into canines or caniniform teeth; 1 to 3 well developed canines near midlength of dentary. Vomer and palatines with teeth; vomerine teeth in chevron-shaped patch without posterior prolongation. Endopterygoids and tongue toothless.

Single dorsal fin (not divided to base at junction of spinous and soft-rayed portions, but fin may appear notched at junction). Pectoral fin essentially symmetrical. Caudal fin truncate to slightly emarginate; principal rays 17 (9 + 8); branched rays 15 (8 + 7); procurrent rays 5 to 10 dorsally, 5 to 9 ventrally. Vertebrae 26 (10 + 16). First caudal vertebra without parapophyses. Formula for configuration of supraneural bones, etc. 0/0+0/2/1+1/1/ (Fig. 2A). Pleural ribs on vertebrae 3 through 10. Epineurals associated with first 12 or 13 vertebrae. No dorsal trisegmental pterygiophores; anal trisegmental pterygiophores 0 to 3.

Lateral line complete, running parallel to dorsal body contour a few scale rows below dorsal fin, curving to near midlateral axis of body on caudal peduncle. Scales ctenoid. Most of head posterior to anterior margin of orbit with scales; portions of head anterior to orbit variously with or without scales (see individual species accounts); maxilla usually naked (occasionally with a few scales). Basal portions of fins scaly for variable distances distally. Scaly interpelvic process between bases of pelvic fins well developed.

Remarks. The synonymy presented is almost entirely from Randall (1980). Randall (1980) revised (but did not demonstrate the monophyly of) *Plectranthias*, subsuming eight other genera into its synonymy while recognizing 30 species, 13 of them new, as members of the genus. Since Randall's revision, 18 other species have been described as new and assigned to *Plectranthias* (Randall, 1996; Zajonz, 2006; Anderson, 2008; Heemstra and Randall, 2009; Wu et al., 2011). Additionally, there are other undescribed species that in the present system of classification would be assigned to *Plectranthias* (Randall and Hoese, 1995; Randall, 1996). It seems likely that future work will demonstrate the validity of one or more of the genera placed in the synonymy of *Plectranthias* by Randall (1980). Because two different types of scales are present in the species now relegated to the genus *Plectranthias* and because type of scale appears to be important in anthiine classification (Anderson et al., 1990), we anticipate the ultimate placement of the western Atlantic *P. garrupellus*, with the presumed derived type of scale, in a genus distinct from that containing the three apparently closely related eastern Pacific species, *P. exsul*, *P. nazcae*, and *P. parini*, all with the primitive type of scale (Anderson and Randall, 1991; Anderson, 2008). Randall (1980:102) considered *Ellerkeldia* (= *Hypoplectrodes*) to be "closely related to *Plectranthias*." For remarks on this alleged relationship, see **Comments on the taxonomy of Hypoplectrodes** under the genus *Hypoplectrodes*.

Key to the Atlantic and Eastern Pacific Species of *Plectranthias*

1a. Total gillrakers on first arch 14–18; lateral-line scales 28 or 29; pectoral-fin rays 12–14 (usually 13); circum-caudal-peduncular scales 12–14 (usually 14); scales ctenoid, with only marginal cteni (i.e., without ctenial bases present proximal to marginal cteni; Fig. 1B); ventral border of preopercle with 1–3 (usually 2) antrorse spines..................... *Plectranthias garrupellus* (western North Atlantic)

1b. Total gillrakers on first arch 26–31; lateral-line scales 36–46; pectoral-fin rays 15–17 (usually 16); circum-caudal-peduncular scales 16–22; scales ctenoid, with rows of ctenial bases present proximal to marginal cteni (Fig. 1A); ventral border of preopercle usually without antrorse spines 2

2a. Scales on dorsum of head not extending anterior to nares (Randall, 1996:117); longest soft dorsal-fin ray 26–35% SL; total gillrakers on first arch 26–28; in life, two orange-red bars on body (one almost entirely anterior to anal-fin base, the other terminating ventrally posterior to anal-fin base) .... *Plectranthias parini* (eastern South Pacific)

2b. Scales on dorsum of head extending anteriorly almost to upper lip, except for triangular premaxillary groove (Randall, 1996:117); longest soft dorsal-fin ray 16–26% SL; total gillrakers on first arch 26–31 (usually 28 or 29); coloration not as described in 2a 3

3a. Circum-caudal-peduncular scales 17 or 18; tubed lateral-line scales 36–42 (mean 39.7); total gillrakers on first arch 28–31 (mean 28.8)......................... *Plectranthias nazcae* (eastern South Pacific)

3b. Circum-caudal-peduncular scales 20–22; tubed lateral-line scales 40–46 (mean 43.0); total gillrakers on first arch 26–29 (mean 27.8)......................... *Plectranthias exsul* (eastern South Pacific)

Plectranthias exsul Heemstra and Anderson, 1983

Exiled Basslet

Fig. 7; Tables 2–8; Map 3

Plectranthias exsul Heemstra and Anderson, 1983:632, figs. 1, 2 (original description, illustrations; holotype ANSP 127843, 158 mm SL; type locality off Juan Fernández Islands; 33°37'S, 78°49'W).—Meléndez and Villalba, 1992:10, fig. 15 (description, illustration, Juan Fernández Islands).—Rojas and Pequeño, 1998a:185, fig. 7 (description, illustration; Juan Fernández and Desventuradas islands).—Anderson, 2008:435–436 (coloration, distribution; in key).

Plectranthias lamillai Rojas and Pequeño, 1998b:205, figs. 2, 3 (original description, illustrations; holotype MNHNC P. 7055, 136 mm SL; type locality off Alejandro Selkirk Island, Juan Fernández Islands).—Rojas and Pequeño, 1998a:187, fig. 8 (description, illustration; Alejandro Selkirk Island, Juan Fernández Islands).—Rojas and Pequeño, 1998c:48, fig. 4, table 3 (description, illustration; Alejandro Selkirk Island, Juan Fernández Islands).—Pequeño and Sáez, 2000:32 (endemic to Juan Fernández Islands).—Anderson and Baldwin, 2000:380 (considered as junior synonym of *P. exsul*).—Anderson and Baldwin, 2002:233; figs. 1, 2; tables 1, 2 (placed in synonymy of *P. exsul*).

Diagnosis. A species of *Plectranthias* distinguishable from all other members of the genus, except *P. nazcae* and *P. parini*, by number of gillrakers (total on first arch 26–29 in *P. exsul* vs. 13–25, usually fewer than 20). It is separable from those two species by numbers of circum-caudal-peduncular scales (20–22 in *P. exsul* vs. 16–18) and tubed lateral-line scales (40–46, mean 43.0 in *P. exsul*, vs. 36–42, mean 39.7 in *P. nazcae*, and 37–40, mean 38.2 in *P. parini*) and by certain body proportions (see Table 8).

Description. Dorsal-fin rays X, 15 or 16. Anal-fin rays III, 7. Pectoral-fin rays 16 or 17. Gillrakers on first arch 8 or 9 + 18 to 20–total 26 to 29. Tubed lateral-line scales 40 to 46. Circum-caudal-peduncular scales 20 to 22.

Preopercle with or without antrorse spines on horizontal limb, no spine at angle, both vertical and horizontal limbs serrate. Scales ctenoid, with rows of ctenial bases present proximal to marginal cteni (Fig. 1A). Most of head scaly. Interorbital region, dorsum of snout (except most anterior part and middorsal area), and posterior one-third or more of ventral surface of lower jaw covered with scales. Maxilla naked or with very few scales; lateral aspect of snout, lachrymal, most of lower jaw, gular region, branchiostegals, and branchiostegal membranes without scales; gular region occasionally with scales anteriorly. Posterior part of spinous dorsal fin with scales basally; soft dorsal and anal fins with scales for some distance distally on fins; pectoral, pelvic, and caudal fins scaly basally; no enlarged axillary scales at base of pelvic fin.

Internarial distance 10 to 17 times in snout length. Head length 38 to 40% SL. Snout length 9 to 12% SL. Orbit diameter 8 to 10% SL. Body depth at first dorsal spine 34 to 40% SL. Longest dorsal spine fourth, fifth, or sixth, 18 to 20% SL. Longest dorsal soft ray the second, produced slightly, 21 to 26% SL. Depressed anal-fin length 27 to 31% SL. Pelvic-fin length 23 to 26% SL. Caudal fin truncate to slightly concave, with one of dorsalmost branched rays elongated. Upper caudal-fin lobe 28 to 30% SL. Lower caudal-fin lobe 24 to 26% SL. Heemstra and Anderson (1983) gave a detailed description of *P. exsul*.

Coloration. Anderson and Baldwin (2000, 2002) examined the holotype and only known specimen of *Plectranthias lamillai* Rojas and Pequeño, 1998b, and synonymized that species with *P. exsul* Heemstra and Anderson, 1983. Rojas and Pequeño (1998b:207) described the fresh coloration of the holotype (MNHNC P. 7055, 136 mm SL) of *Plectranthias lamillai* as:

> Body pale yellowish, with a broad red bar from sixth dorsal-fin spine to base of fifth ray of dorsal fin, extending to anus and above anal fin as a narrow band that widens on the peduncle and then bifurcates over the upper and lower margins of the caudal fin; rest of the caudal fin yellowish. An irregular red blotch on nape and below first three dorsal-fin spines, extending to opercle, subopercle, cheek, snout and front of upper jaw; maxilla yellow; lower jaw reddish. Rim of pupil yellow; iris dark with reddish spots. Pectoral fins orange; pelvic an[d] anal fins whitish.

Anderson and Baldwin (2000, 2002) considered the coloration of *P. lamillai* as illustrated and described in the original description (Rojas and Pequeño 1998b:205

[fig. 2], 207) to be a variation on the pattern displayed by *P. exsul* because they had mistakenly considered a color transparency of *Plectranthias nazcae* to be of *P. exsul*. Despite this confusion, Anderson and Baldwin (2000, 2002) were correct in ascribing the pattern of coloration presented by Rojas and Pequeño (1998b) for *P. lamillai* to be of *P. exsul*.

Rojas and Pequeño (1998a:186) described the fresh coloration of *Plectranthias exsul*, based on material identified by them and apparently collected off Isla Robinson Crusoe (Juan Fernández Islands), as:

> *Color: especímenes frescos presentan la región dorsal cefálica anaranjada a rojiza. El área preopercular cerca de la cabeza anaranjada. El cuerpo rojizo, con una amplia banda roja que se extiende desde la base posterior de la aleta dorsal blanda hasta el pedúnculo caudal. Las aletas son amarillentas, con manchas anaranjadas sobre la porción dorsal espinosa. Iris amarillento.*

> Color: fresh specimens with dorsal cephalic region orange to reddish. Preopercular region of head orange. Body reddish with broad red band extending from posterior base of soft dorsal fin to caudal peduncle. Fins yellowish, with orange spots on spinous dorsal fin. Iris yellowish. (Translation by WDA.)

The coloration described by Rojas and Pequeño (1998a:186) for *P. exsul* is similar to that displayed by *P. nazcae* (see Fig. 9). We have not examined the material on which Rojas and Pequeño (1998a) based their account nor have we seen color photographs of those specimens.

Meléndez and Villalba (1992:10) provided a brief description of a specimen (MNHNC P. 6766, 144 mm SL) that was collected in the Juan Fernández Islands and identified as *Plectranthias exsul*. Meristic and morphometric data in their description agree well with data for those characters in *P. exsul*. They described the coloration as *cuerpo anaranjado claro* (= body light orange). Apparently the coloration of *P. exsul* is quite variable.

Plectranthias lamillai. Rojas and Pequeño (1998b) described a new species of *Plectranthias*, *P. lamillai*, from a single specimen (MNHNC P.7055, 139.6 mm SL, now 136 mm SL) collected off Alejandro Selkirk Island in the Juan Fernández Islands off the coast of Chile. In addition to their original description, Rojas and Pequeño (1998a, c) provided accounts of *P. lamillai* in two other papers, and Pequeño and Saez (2000) referred to it as a species endemic to Juan Fernández. Anderson and Baldwin (2000, 2002) compared the holotype of *P. lamillai* with type material of *P. exsul* and considered the differences noted by Rojas and Pequeño (1998b) as insufficient to justify the recognition of *P. lamillai* and *P. exsul* as distinct species. We follow Anderson and Baldwin and consider *P. lamillai* to be a junior synonym of *P. exsul*.

Remarks. Differences in body proportions shown in Table 8 may be reflections of the small sample sizes rather than real differences among the species. The specimen

(USNM 312927, 122 mm SL) depicted in the center of fig. 1 in Anderson and Baldwin (2002:236) and identified as *P. exsul* is a paratype of *P. nazcae*.

Distribution. We have examined specimens of *P. exsul* from the Juan Fernández Islands off the coast of Chile. Depths of capture are available for two of the lots examined: 140 to 165 meters and 180 to 200 meters. Meléndez and Villalba (1992) reported a specimen of 144 mm SL collected from a depth of only 8 meters.

Rojas and Pequeño (1998a) included the Islas Desventuradas in the range of *P. exsul* but did not list any material examined from those islands. Pequeño and Lamilla (1996a, 1996b, 2000) published three papers on the fishes of the Desventuradas Islands but made no mention of *P. exsul* in any of them. Although it would not be surprising to find that the range of *P. exsul* extends to the Desventuradas, in view of the relatively close proximity of those islands to the Juan Fernández group which is about 750 kilometers to the south, occurrence of that species at the Desventuradas needs verifying.

Material examined. Six specimens, 134 to 158 mm SL. **JUAN FERNÁNDEZ ISLANDS:** ANSP 127843 (holotype: 158 mm SL), MCZ 52520 (2: 134–140), MNHNC P.5611 (1: 154), MNHNC P.7055 (holotype of *Plectranthias lamillai*: 136), USNM 176577 (1: 158).

Plectranthias garrupellus Robins and Starck, 1961

Apricot Bass

Fig. 8; Tables 2–7; Map 4

Plectranthias garrupellus Robins and Starck, 1961:295, fig. 7—middle (original description, illustration; holotype ANSP 95129, 52 mm SL; type locality off Daytona Beach, Florida—28°52′N, 80°05′W in 119 meters).—Anderson, 2003:1363 (species account).

Diagnosis. A species of *Plectranthias* distinguishable from the other species of the genus included herein by the following combination of characters. Scales ctenoid, with only marginal cteni, no ctenial bases in posterior fields (Fig. 1B). Total gillrakers on first arch 14 to 18; lateral-line scales 28 or 29; pectoral-fin rays 12 to 14 (usually 13); circum-caudal-peduncular scales 12 to 14 (usually 14); ventral border of preopercle with 1 to 3 (usually 2) antrorse spines.

Description. Dorsal-fin rays X, 15 to 17 (rarely 17). Anal-fin rays III, 7. Gillrakers on first arch 4 to 6 + 10 to 13—total 14 to 18. Tubed lateral-line scales 28 or 29 (most frequently 29).

Ventral border of preopercle with 1 to 3 (usually 2) antrorse spines, posterior border serrate, no spine at angle. Most of head, including interorbital region, scaly. Maxilla naked or with a few embedded scales that are difficult to see. Snout, lachrymal, lower jaw, gular region, branchiostegals, and branchiostegal membranes naked, except a few scales frequently present on posterior part of lower jaw and central branchiostegal rays often with numerous scales. Bases of all fins heavily covered with scales.

Internarial distance 7 to 14 times in snout length. Head length 40 to 44% SL. Snout length 7 to 11% SL. Orbit diameter 8 to 12% SL. Body depth at first dorsal spine 29 to 42% SL (29–38% SL in specimens 25–49 mm SL, 31–42% SL in specimens 51–73 mm SL). Longest dorsal spine usually the third, occasionally the fourth, 15 to 26% SL. Longest dorsal soft ray 13 to 17% SL. Depressed anal-fin length 27 to 37% SL. Pelvic-fin length 22 to 30% SL. Caudal fin truncate or nearly so. Upper caudal-fin lobe 22 to 29% SL. Lower caudal-fin lobe 22 to 29% SL. Anderson (2003) presented a brief account of *Plectranthias garrupellus*.

Coloration. Bullock and Smith (1991:209, pl. II, fig. D) presented a color photograph of a specimen of ca. 50 mm of *P. garrupellus* and (p. 28) stated: "Overall color red; each scale with underlying basal dark red blotch."

The following description is based on examination of a color transparency provided by Donald D. Flescher of a specimen of *P. garrupellus* (GMBL 81–149, 44 mm SL) caught off Cape Fear, North Carolina. Dorsum of head and body red orange; lateral and ventral parts of head and body mostly rosy; considerable dull yellow orange on cheek; lateral aspect of body with numerous flecks of dull yellow orange. Iris mostly rosy. Spinous dorsal fin red orange; soft dorsal, anal, and caudal fins yellow; pelvic fin mostly pallid, but with some dull yellow.

Sexuality and sexual dimorphism. The only specimens that have been sexed are those examined by Robins and Starck (1961). The three females they reported were 48 to 52 mm SL, whereas the two males were 52 and 61 mm SL, supporting their (Robins and Starck, 1961: 296) contention that *Plectranthias garrupellus* is probably protogynous. We have examined one of the male paratypes (ANSP 94416, now 51 mm SL) and all three females (ANSP 95129, the holotype, 52 mm SL; UF 204523, formerly UMML 4523, now 47 mm SL, and USNM 197392, 49 mm SL, both paratypes) reported by Robins and Starck (1961:296) and have found that the male paratype has relatively greater lengths of certain body parts. On the male depth of body at first dorsal spine is 42% SL, on the females 34 to 37% SL; on the male pelvic-fin length is 30% SL, on the females 22 to 23% SL; and on the male longest dorsal spine is 26% SL, on the females 15 to 18% SL. Another specimen (GMBL 61–22, 53 mm SL), for which the sex has not been determined, shows the following proportions: depth of body at first dorsal spine 39% SL, pelvic-fin length 30% SL, and longest dorsal spine 24% SL. Based on data available for specimens of known sex, the specimen from GMBL 61–22 is presumably a male.

Reproduction. Females with vitellogenic oocytes have been taken off both the east and west coasts of Florida during August–September (Robins and Starck, 1961; Bullock and Smith, 1991).

Early life history. Baldwin (1990:939; fig. 18) provided a description of the early developmental stages of *P. garrupellus* based on three specimens, 5.5 to 8.2 mm SL, and included illustrations of three larvae (two of which were taken from Kendall, 1979), and Richards et al. (2006) presented an account of the early life history of the species.

Ecological notes. Bullock and Smith (1991:28) mentioned that the stomachs of several specimens of *P. garrupellus* from the Gulf of Mexico, west of Cape Sable, Florida, contained "an unidentified shrimp, the crab *Munida* sp., an unidentified xanthid crab, and a worm tube." Gutherz et al. (1995: table 2) observed small numbers of *P. garrupellus*, in depths of 185 to 220 meters, from a submersible operating off Charleston.

Distribution. We have examined specimens collected off the Atlantic coast of the United States (North Carolina—as far north as latitude 33°13′N, South Carolina, Florida), in the Straits of Florida, in the Bahamas, off Cuba (La Habana), in the Caribbean Sea (Mexico—off Quintana Roo, Nicaragua, Dominican Republic, Windward Islands— off Grenada) in depths of 13 to 375 (225/375) meters. Houde (1982) reported the larvae of *Plectranthias garrupellus* from the eastern Gulf of Mexico, and Bullock and Smith (1991) examined 16 specimens (19–53 mm SL) from the southeastern Gulf of Mexico. Claro et al. (2000) noted a single specimen (74 mm SL) captured off the southern coast of Cuba in 293 meters. Roa-Varon et al. (2003) reported two specimens of this species (42–70 mm total length) captured off the Caribbean coast of Colombia in 55 to 210 meters.

Material examined. Forty-five specimens, 25 to 73 mm SL. **NORTH CAROLINA:** GMBL 81–149 (1 specimen: 44 mm SL). **SOUTH CAROLINA:** GMBL 81–130 (1: 26). **FLORIDA (ATLANTIC):** ANSP 95129 (holotype: 52), GMBL 61–20 (1: 46), GMBL 61–21 (2: 46–55), GMBL 61–37 (1: 28), GMBL 62–43 (1: 43), SAIAB 43364 (2: 31–43), UF 15644 (1: 58), UF 204523 (paratype: 47), UF 211623 (1: 44), UF 215388 (2: 25–45), USNM 179244 (1: 35), USNM 197392 (paratype: 49), USNM 326410 (1: 49). **STRAITS OF FLORIDA:** FMNH 70703 (1 of 2: 54). **BAHAMAS:** ANSP 94416 (paratype: 51), FMNH 70704 (1: 49), GMBL 61–22 (1: 53). **CUBA:** MCZ 28048 (1: 61). **MEXICO (QUINTANA ROO):** UF 215283 (1: 44), UF 223665 (1 of 2: 54), UF 227306 (2: 43–44), UF 228455 (3: 44–72), UF 228456 (1: 52), UF 228608 (2: 55–69), UF 228614 (5 of 11: 45–57), UF 228657 (1: 48), UF 229353 (1: 46), UF 229691 (1: 49), USNM 148348 (1: 58). **DOMINICAN REPUBLIC:** UF 229870 (1: 46). **NICARAGUA (CARIBBEAN):** UF 227305 (1: 73). **WINDWARD ISLANDS (GRENADA):** UF 224273 (1: 63).

Plectranthias nazcae Anderson, 2008

Red Splodge

Fig. 9; Tables 2–8; Map 3

Plectranthias sp., Parin et al., 1981:14 (brief description).

Plectranthias exsul (non Heemstra and Anderson, 1983): Heemstra and Anderson, 1983:632, 634, 636 (Nazca Ridge).—Anderson and Randall, 1991:338, fig. 2C; 339–342 (Nazca Ridge).—Parin, 1991:673, 679 (region of Nazca and Sala y Gómez ridges).—Parin et al., 1997:172 (near SW end of Nazca Ridge).— Anderson and Baldwin, 2000:380 (Nazca Ridge).—Anderson and Baldwin, 2002:236, fig. 1—center; 237 (Nazca Ridge).

Plectranthias nazcae Anderson, 2008:430, figs. 1, 2 (original description, illustrations; holotype USNM 312926, 115 mm SL; type locality near SW end of Nazca Ridge in the eastern South Pacific—25°43'S, 85°22'W in 185 to 200 meters).

Diagnosis. A species of *Plectranthias* distinguishable from all other members of the genus, except *P. exsul* and *P. parini*, by number of gillrakers (total on first arch 28–31 in *P. nazcae* vs. 13–25, usually fewer than 20). It is separable from *P. exsul* by numbers of circum-caudal-peduncular scales (17 or 18 vs. 20–22 in *P. exsul*) and tubed lateral-line scales (36–42, mean 39.7; vs. 40–46, mean 43.0, in *P. exsul*) and by certain body proportions (see Table 8). *Plectranthias nazcae* is distinguished from *P. parini* by coloration (*P. nazcae* with red oblong area beneath posterior end of soft dorsal fin that extends onto caudal peduncle vs. *P. parini* with two orange-red bars on posterior half of body) and by certain body proportions, most strikingly in length of longest dorsal soft ray and depth of caudal peduncle (see Table 8).

Description. Dorsal-fin rays X, 16. Anal-fin rays III, 7. Pectoral-fin rays 16 or 17. Gillrakers on first arch 7 to 10 + 19 to 22—total 28 to 31. Preopercle without antrorse spines on horizontal limb, no spine at angle, both vertical and horizontal limbs serrate. Scales ctenoid, with rows of ctenial bases present proximal to marginal cteni (Fig. 1A). Most of head covered with scales. Interorbital region, usually dorsum of snout (except most anterior part and middorsal area), and usually posterior one-third to two-thirds of ventral surface of lower jaw covered with scales. Maxilla (usually), lateral aspect of snout, lachrymal, gular region, branchiostegals, and branchiostegal membranes without scales. Posterior part of spinous dorsal fin with scales basally; soft dorsal and anal fins with scales for some distance distally on fins; pectoral, pelvic, and caudal fins scaly basally; no enlarged axillary scales at base of pelvic fin.

Internarial distance 9 to 12 times in snout length. Head length 37 to 39% SL. Snout length 8 to 11% SL. Orbit diameter 8 to 11% SL. Body depth at first dorsal spine 30 to 36% SL. Longest dorsal spine fourth or fifth, 16 to 19% SL. Longest dorsal soft ray the second, produced slightly, 16 to 23% SL. Depressed anal-fin length 26 to 30% SL. Pelvic-fin length 22 to 25% SL. Caudal fin truncate to slightly emarginate with upper lobe usually slightly produced. Upper caudal-fin lobe 24 to >30% SL. Lower caudal-fin lobe 22 to >27% SL. Anderson, 2008, gave a detailed description of *P. nazcae*.

Coloration. Anderson and Randall (1991:341) mistakenly attributed a color transparency taken shortly after capture of a specimen of *Plectranthias nazcae* to *P. exsul* and provided a description (p. 341) of the coloration of that specimen and a black and white illustration (p. 338, fig. 2C) of it. Their description follows:

> Head dark red-orange dorsally, with somewhat diffuse brighter red-orange area on preopercle near orbit; posterior and ventral parts of head lighter—mostly dull yellow. Iris dull yellow anteriorly and posteriorly, the yellow bordered by a narrow red-orange rim anteriorly, suffused with red-orange posteriorly; dorsal and ventral portions of iris largely melanistic. Body generally with reddish-orange cast, darker dorsally; a well-defined brilliant red oblong area extending from bases of posterior dorsal soft rays ventrally

to just below midline and then posteriorly over middle of caudal peduncle to reach mid-ventral line near base of caudal fin (see pattern of coloration in Fig. 2C). Fins with dull yellowish cast (some dull orange on spinous portion of dorsal fin), but without any distinctive pigmentation.

Remarks. Differences in body proportions shown in Table 8 may be reflections of the small sample sizes rather than real differences among the species. The specimen (USNM 312927, 122 mm SL) depicted in the center of fig. 1 in Anderson and Baldwin (2002:236) and identified as *P. exsul* is a paratype of *P. nazcae*.

Distribution. *Plectranthias nazcae* is known only from specimens collected near the southwest end of the Nazca Ridge in the eastern South Pacific, some 1500 kilometers west of Chile, at about 26°S, in depths of 162 (162/168) to 225 (200/225) meters. It is the only species of *Plectranthias* recorded from the Nazca Ridge.

Material examined. Five specimens, 37.6 to 150 mm SL. **NEAR SOUTHWEST END OF NAZCA RIDGE:** BPBM 27978 (1 specimen: 133 mm SL), BPBM 29400 (1: 37.6), USNM 312926 (holotype: 115), USNM 312927 (1: 122), ZMMU P-16022 (1: 150).

Plectranthias parini Anderson and Randall, 1991

Twicebarred Perch

Fig. 10; Tables 2–8; Map 3

Plectranthias parini Anderson and Randall, 1991:336, figs. 1, 2 (original description, illustrations; holotype USNM 312925, 85 mm SL; type locality Sala y Gómez Ridge in the eastern South Pacific; 25°02.6′S, 97°29.2′W in 260 to 272 meters).— Randall, 1996:128, fig. 3 (range extension to Easter Island).—Rojas and Pequeño, 1998a:189, fig. 9 (description, illustration—adapted from Randall, 1996).— Anderson, 2008:436 (in key).

Diagnosis. A species of *Plectranthias* distinguishable from all other members of the genus, except *P. exsul* and *P. nazcae*, by number of gillrakers (total on first arch 26–28 in *P. parini* vs. 13–25, usually fewer than 20). It is separable from *P. exsul* by numbers of circum-caudal-peduncular scales (16 or 17 vs. 20–22 in *P. exsul*) and tubed lateral-line scales (37–40, mean 38.2, vs. 40–46, mean 43.0, in *P. exsul*) and by certain body proportions (see Table 8). *Plectranthias parini* is distinguished from *P. nazcae* by coloration (*P. parini* with two orange-red bars on posterior half of body vs. *P. nazcae* with red oblong area beneath posterior end of soft dorsal fin that extends onto caudal peduncle) and by certain body proportions, most strikingly in length of longest dorsal soft ray and depth of caudal peduncle (see Table 8).

Description. Dorsal-fin rays X, 16. Anal-fin rays III, 7. Pectoral-fin rays 15 or 16. Gillrakers on first arch 8 + 18 to 20—total 26 to 28. Preopercle with both free margins serrate, without antrorse spines or spine at angle. Scales ctenoid, with rows of ctenial bases present proximal to marginal cteni (Fig. 1A). Most of head covered with scales. Interorbital region heavily covered with scales; posterior part of ventral surface of lower jaw with scales. Most of dorsum of snout, lateral aspect of snout, maxilla, supramaxilla, lachrymal, almost all of suborbital region, most of lower jaw, gular region, branchiostegals, and branchiostegal membranes without scales. Posterior part of spinous dorsal fin scaly basally; basal one-third to one-half of soft dorsal and anal fins scaly; pectoral, pelvic, and caudal fins scaly basally; pelvic axillary process poorly developed. Internarial distance 7 to 13 times in snout length. Head length 39 to 41% SL. Snout length 8 to 12% SL. Orbit diameter 9 to 12% SL. Body depth at first dorsal spine 36 to 38% SL. Longest dorsal spine the fifth, 15 to 18% SL. Longest dorsal soft ray the second, produced, 26 to 35% SL. Depressed anal-fin length 29 to 34% SL. Pelvic fin length 26 to 29% SL. Caudal fin truncate except two dorsalmost branched rays produced in holotype and dorsalmost and ventralmost branched rays produced in the other specimen examined. Upper caudal-fin lobe 36% SL in holotype. Lower caudal-fin lobe 27% SL in holotype. (Caudal lobes damaged distally on the other specimen examined.) Anderson and Randall (1991) provided a detailed description of this species.

Coloration. The following description was made from a transparency taken shortly after capture of the holotype (Anderson and Randall, 1991:338).

> Head mainly pale orange, dull yellow ventrally. Iris dull yellow except for a hint of reddish coloration on posterior border. Body with an overall orange cast dorsally, yellowish ventrally. Two orange-red parallel bars on body; first originating well out on interradial membranes 5 through 9 of spinous portion of dorsal fin and extending slightly obliquely to terminate ventrally almost entirely anterior to first anal spine—first bar slightly broader at dorsal-fin base than at ventral termination; second bar beginning on bases of posterior dorsal soft rays and on dorsum of anterior part of caudal peduncle and extending slightly obliquely to terminate ventrally posterior to anal-fin base. . . . Except where orange-red bars encroach on dorsal fin, fins mostly pallid, but anteriormost interradial membranes of spinous dorsal fin with some yellow-orange.

Anderson and Randall (1991, figs. 1, 2A), Randall (1996, fig. 3), and Rojas and Pequeño (1998a, fig. 9) provided illustrations of this species that clearly show the locations of the orange-red bars.

Remark. Differences in body proportions shown in Table 8 may be reflections of the small sample sizes rather than real differences among species.

Distribution. *Plectranthias parini*, known only from the eastern South Pacific, has been collected on the Sala y Gómez Ridge (Anderson and Randall, 1991; Parin, 1991; Parin et al., 1997) and off Easter Island (Randall, 1996) in depths of 229 to 272 (260/272) meters. The locality nearer South America is about 2700 kilometers west of Chile.

Material examined. Two specimens, 84.7 to 163 mm SL. **SALA Y GÓMEZ RIDGE:** USNM 312925 (holotype: 84.7 mm SL). **EASTER ISLAND:** BPBM 33460 (1: 163).

HEMANTHIAS STEINDACHNER, 1875

Table 1

Hemanthias Steindachner, 1875:378, proposed as a subgenus of *Anthias* (type species *Anthias (Hemanthias) peruanus* Steindachner, 1875:378, by monotypy).

Centristhmus Garman, 1899:47 (type species *Centristhmus signifer* Garman, 1899:48, by monotypy).

Anthiasicus Ginsburg, 1952:91 (type species *Anthiasicus leptus* Ginsburg, 1952:91, by original designation).

Diagnosis. *Hemanthias* is distinguishable from all other genera of Anthiinae covered in this work by the following combination of characters. Dorsal-fin rays X, 12 to 15 (usually 14). Anal-fin rays III, 7 to 9 (usually 8). Gillrakers well developed, total number on first arch 31 to 39 (usually 32–37). Lateral line complete, not interrupted, tubed scales 48 to 71 (usually 51–67). Circum-caudal-peduncular scales 34 to 50. Vomerine tooth patch chevron shaped without posterior prolongation. Preopercle serrate but without antrorse spines on ventral border. Urohyal with or without anteriorly projecting spine. Scales ctenoid, with only marginal cteni, no ctenial bases in posterior fields (Fig. 1B). No secondary squamation. Maxilla without scales.

Baldwin (1990) noted three characters that appear to be synapomorphies for *Hemanthias* and *Choranthias*. They are: larvae with internal patch of pigment on posterior part of trunk between dorsal midline and midbody or at midbody (observed in *H. leptus, H. signifer,* and *C. tenuis*); larvae, at some time during development, with small knob or low smooth ridge on supraoccipital (observed in *H. leptus, H. signifer,* and *C. tenuis*); and parapophyses present on first caudal (i.e., the eleventh) vertebra. (Examination of radiographs of three specimens of *C. salmopunctatus* reveals that two appear to possess parapophyses on the first caudal vertebra, but the third may lack them.)

Description. With mouth closed, lower jaw exceeding upper jaw. Premaxillae protrusile. No supramaxilla. Anterior and posterior nares rather closely set on each side of snout; posterior border of anterior naris produced into a short flap but never into a long filament. No fleshy papillae on border of orbit. Upper limb of preopercle with mostly small serrae; lower limb without antrorse spines, usually with larger serrae than on upper limb; angle of preopercle with single well-developed spine, one to several spines or spine-like processes, or enlarged serrae.

Premaxilla with inner band of very small teeth and outer series of larger conical teeth; near symphysis, usually one to several teeth along medial margin of inner band enlarged as posteriorly directed conical teeth; outer row of larger conical teeth usually preceded by a larger canine or canine-like tooth. Dentary with row of conical teeth along lateral edge of jaw; this row including one or two teeth, usually enlarged into recurved canines, at a point approximately one third length of row from its anterior end; band or row of very small teeth extending anteriorly from this row and reaching to near symphysis; usually one or two teeth on inner edge of band near symphysis enlarged and directed posteriorly; usually one or two enlarged exserted canine or canine-like teeth near symphysis. Vomer and palatines with teeth; teeth on vomer in chevron-shaped patch, without posterior prolongation, teeth at apex and posterior corners of vomerine tooth patch usually enlarged; palatine teeth small, in single row or narrow band; no teeth on endopterygoids or tongue.

Single dorsal fin; not deeply notched at junction of spinous and soft-rayed parts. Pectoral fin roughly symmetrical with 15 to 21 rays. Principal caudal-fin rays 15 (8 + 7); branched rays 13 (7 + 6). Vertebrae 26 (10 + 16). Formula for configuration of supraneural bones, etc. 0/0/2/1+1/1/ (Fig. 2B). Pleural ribs on vertebrae 3 through 10. Epineurals associated with first 10 to 12 vertebrae. No trisegmental pterygiophores associated with dorsal and anal fins.

Lateral line complete, extending to at least base of caudal fin (curving upward beneath spinous dorsal fin, then descending gradually to course near midlateral part of caudal peduncle). Much of head covered with scales, but snout, lachrymal region, maxilla, anterior part to most of interorbital, gular region, branchiostegals, and branchiostegal membranes without scales; lower jaw mostly without scales (scales posteriorly in *H. signifer* and occasionally in *H. peruanus*). Dorsal and anal fins mostly without scales, although soft parts of those fins may be more or less scaly basally; pectoral, pelvic, and caudal fins scaly basally. Modified scales (interpelvic process) overlapping pelvic-fin bases along midventral line.

Key to the Species of *Hemanthias*

1a. Posterior 50–70% of lower jaw covered with scales; lateral-line scales 60–71 (usually 62–68); specimens greater than about 70 mm SL with exposed spine projecting anteriorly from ventral border of urohyal . . . .*Hemanthias signifer* (eastern Pacific)

1b. Lower jaw usually without scales (occasionally with some scales posteriorly in *H. peruanus*); lateral-line scales 48–63 (usually 50–61); urohyal without anteriorly projecting spine 2

2a. Sum of lateral-line scales plus total gillrakers on first arch for individual specimens 90–102 (usually 92–98); total gillrakers on first arch 34–39; lateral-line scales 54–63; in specimens less than about 220 mm SL caudal fin deeply forked, in those greater than about 310 mm SL caudal fin almost truncate . . . . *Hemanthias leptus* (western Atlantic)

2b. Sum of lateral-line scales plus total gillrakers on first arch for individual specimens 81–89 (usually 83–88); total gillrakers on first arch 31–35; lateral-line scales 48–59; caudal fin forked with middle rays of each caudal-fin lobe longest in specimens greater than about 100 mm SL . . . . *Hemanthias peruanus* (eastern Pacific)

Hemanthias leptus (Ginsburg, 1952)

Longspine Bass

Fig. 11; Tables 2–7; Map 4

Anthiasicus leptus Ginsburg, 1952:87, 91, fig. 4 (original description; illustration; holotype USNM 134189, 108 mm SL; type locality off Dauphin Island, Alabama).—Ginsburg, 1954:263, 264, figs. 5 & 6 (remarks; two illustrations: one of holotype, one of USNM 157788).

Diagnosis. A species of *Hemanthias* distinguishable from the other members of the genus by the following combination of characters. Lower jaw without scales. Sum of total gillrakers on first arch plus number of tubed lateral-line scales in individual specimens 90 to 102 (usually 92–98, mean 95.1). Lateral-line scales 54 to 63 (usually 55–61, mean 58.1). Urohyal without anteriorly projecting spine. In specimens less than about 220 mm SL caudal fin deeply forked, in those greater than about 310 mm SL caudal fin almost truncate.

Description. Dorsal-fin rays X, 13 or 14 (usually 14). Anal-fin rays III, 7 or 8 (usually 8). Pectoral-fin rays 17 to 20 (usually 18, rarely 17 or 20). Gillrakers 9 to 12 + 24 to 29—total 34 to 39. Circum-caudal-peduncular scales 40 to 46 (one specimen with 36). Procurrent caudal-fin rays 13 or 14 dorsally, 12 to 14 ventrally. Epineurals associated with first 10 or 11 vertebrae.

Internarial distance 7 to 13 times in snout length. Head length 30 to 36% SL. Snout length 5 to 11% SL (7–11% SL in specimens more than 80 mm SL). Orbit diameter 6 to 13% SL (6–9% SL in specimens more than 260 mm SL). Body depth at first dorsal spine 27 to 38% SL (32–38% SL in specimens more than 240 mm SL). Longest dorsal spine the fourth in specimens less than 120 mm SL; third spine longest in specimens more than about 150 mm SL; longest dorsal spine 10 to >14% SL in specimens 50 to 150 mm SL; in larger individuals third dorsal spine frequently not distinctly separated from its filament; third dorsal spine plus inseparable filament 29 to 46% SL in specimens more than about 200 mm SL. Depressed anal-fin length 32 to 45% SL. Pelvic-fin length 24 to 76% SL (40–76% SL in specimens more than about 180 mm SL). Upper caudal-fin lobe 23 to >55% SL (23–33% SL in specimens more than about 305 mm SL). Lower caudal-fin lobe 24 to 62% SL (24–34% SL in specimens more than about 305 mm SL). Anderson (2003) gave a brief description of the species.

Coloration. Bullock and Smith (1991) presented color photographs of two specimens of *H. leptus*, one of those (p. 207, plate I, fig. D; same photograph in Bullock and Godcharles, 1982, p. 55, fig. 1) being of a small sexually mature male (FSBC 12052, 61 mm SL) and the other (p. 209, plate II, fig. A) of a much larger male (FSBC 12023, 313 mm SL), and wrote (p. 20) :

> General body color carmine; olive freckles; golden stripe below eye from tip of snout to middle of pectoral-fin base; second golden stripe from eye to upper base of pectoral fin; fins yellow with olive freckles. Possible sexual dichromatism noted in small mature male . . . [with] burgundy dorsal-fin stripe and caudal lobes.

William F. Smith-Vaniz (*in litt.* to WDA, 13 March 2000) mentioned the capture of "a very small individual with burgundy dorsal-fin stripe and caudal lobes identical to the specimen photograph (plate I, fig. D) in Bullock and Smith (1991)."

Cervigón (1991:396, fig. 262) published a color photograph of *H. leptus* which shows head to be dull orange dorsolaterally, pink laterally; yellow band from maxilla to posterior end of opercle; body dull orange dorsolaterally, pink laterally, pallid ventrally; dorsal and anal fins mostly pallid proximally, yellow distally; pectoral fin yellow proximally, pink distally; pelvic and caudal fins mostly yellow.

A juvenile specimen (GMBL 09–011, 58 mm SL) examined six days after collection had the following coloration: head yellow dorsally, pallid ventrally; body yellow dorsoposterior to eye, pale yellow immediately ventral to dorsal fin, rest of body pallid; dorsal fin mostly yellow except for stripe on distal border—stripe orange on spinous dorsal and red on soft dorsal; anal fin with a little yellow posteriorly but mostly pallid; pectoral and pelvic fins pallid; caudal fin mostly pallid except dorsal and ventral lobes each with a bright stripe peripherally—stripes yellow proximally grading into bright red distally.

Coloration remaining on a 414-mm SL specimen (GMBL 84–45) 13 days after capture was: head mostly dark rose dorsally, mostly pale rose ventrally, fairly broad bright yellow stripe from maxilla through ventral part of eye to posterior end of opercle—stripe broadening at posterior end of opercle; iris of eye mainly yellow on ventral and dorsoposterior aspects, remainder of iris dark rose; dorsolaterally body rose overlain by numerous dull greenish-yellow flecks, beneath lateral line numerous yellow spots overlying rose posteriorly—blotches of yellow present anteriorly, ventrolaterally body pale rose to pallid (darker dorsally), midventral line from pelvic fins forward to posterior part of isthmus pale yellow; dorsal fin mostly yellow except spinous dorsal with considerable rose at bases of interradial membranes and soft dorsal with rose extending out along soft rays for some distance distally; anal fin mostly yellow except some pale rose at base of fin, yellow pigment intensified into numerous spots on most interradial membranes; pectoral fins yellow basally, pale rose distally; pelvic fins with some rose basally but mainly yellow, elongated rays bright yellow; caudal fin with numerous yellow spots overlying rose.

The following description is based on color transparencies of four adult specimens *of H. leptus*. Unfortunately, the photographs cannot be associated with preserved material. Head rosy dorsally, purplish pink to pallid laterally, rose to pallid ventrally; iris of eye mostly yellow with some reddish, mainly dorsoanteriorly and posteriorly; conspicuous yellow band beginning near anterior end of snout running posteriorly ventral to orbit to end at about posterior end of opercle; body mostly rosy dorsally, pink to pallid laterally (with a scattering of yellow), mostly pallid ventrally; fins with much yellow, mainly distally, and with considerable rose, mostly proximally.

Sexuality. Bullock and Smith (1991:20) found histological evidence for protogyny in *Hemanthias leptus* and, based on histology of an 86-mm ripening male, entertained the idea that *H. leptus* may be diandric. The report of a small sexually mature male (FSBC 12052, 61 mm SL) by Bullock and Godcharles (1982) supports the notion of diandry in this species.

Early life history. Baldwin (1990:934–936; figs. 1, 15) described the early developmental stages of *Hemanthias leptus* based on 17 specimens, 2.0 mm NL to 20.0 mm SL, and included illustrations of four larvae (three of which were taken from Kendall, 1979, who identified them as *Pronotogrammus aureorubens*), and Richards et al. (2006) presented an account of the early life history of the species.

Ontogenetic changes. Ginsburg (1954:264) noted changes with growth in the third dorsal spine and the second pelvic soft ray, both becoming greatly elongated and filamentous in a 240-mm SL specimen as compared with the 108-mm SL holotype, and mentioned that "the most remarkable growth change takes place in the shape of the caudal fin, from being deeply lunate" to almost truncate with a median notch. Bullock and Smith (1991:20) wrote:

Small specimens (<120 mm) with short filaments or tabs at tips of dorsal-fin spines and caudal fin deeply forked. . . . Intermediate-sized individuals (160–270 mm) have moderately elongate third dorsal-spine filament and filamentous pelvic and caudal fins. Sexually transitional specimen (216 mm) with more truncate caudal fin and with both pelvic fin and third dorsal-spine filament extremely elongate. In larger specimens (>270 mm), third dorsal spine with long filament and second pelvic ray elongated; caudal fin basically truncate with medial notch.

Ecological notes. Large schools of very small *Hemanthias leptus* associated with a relatively small number of large individuals of that species have been observed via ROV over the outer continental shelf off Alabama (W. F. Smith-Vaniz, *in litt.* to WDA, 13 March 2000). Bullock and Smith (1991) found that the gut of a 71-mm male *H. leptus* collected in the eastern Gulf of Mexico contained copepods, ostracods, amphipods, and euphausiids and observed fish eyes in the gut contents of another specimen collected off the east coast of Florida. A specimen of *Seriola dumerili* caught at Campeche Bank in the Gulf of Mexico had a specimen of *Hemanthias leptus* (UF 28457, 106 mm SL) taken from its throat.

Distribution. We have examined specimens collected off North Carolina, South Carolina, Florida Keys, west coast of Florida, Alabama, Louisiana, Texas, Mexico (Campeche Bank), Tobago, Guyana, and Suriname in depths of 37 to 640 meters. Reports of this species, as noted below, are from specimens collected or observed off South Carolina, east coast of Florida, Florida Keys, eastern and northwestern Gulf of Mexico, Alabama, Texas, and Venezuela. Briggs et al. (1964) noted a 310-mm specimen taken directly off Port Mansfield, Texas, by hook and line at the 50-fathom contour. Ross et al. (1981) reported a 456-mm specimen (UF 27790; our measurement of SL is 429 mm) of *H. leptus* that was caught by hook and line at a depth of 168 meters off Murrells Inlet, South Carolina. Bullock and Godcharles (1982) mentioned two specimens (FSBC 11807-1, 387 mm SL; FSBC 11807-2, 403 mm SL) of *H. leptus* caught off southwest Florida in the eastern Gulf of Mexico, and Bullock and Smith (1991) reported specimens collected off the east coast of Florida and the eastern and northwestern Gulf of Mexico. Cervigón (1991) provided a description and illustrations (p. 369, description; p. 371, fig. 262, black & white drawing; p. 396, fig. 262, color photograph) of *H. leptus* and reported four specimens (255–310 mm SL, 420–540 mm total length) collected east of Isla La Tortuga, Venezuela. William F. Smith-Vaniz (*in litt.*, 13 March 2000, to WDA) informed us of observations made (via ROV) and of specimens collected on the outer continental shelf off Alabama. On 21 December 2005 WDA received from Ron Martin via e-mail attachment a color photograph of a specimen of *H. leptus* (estimated to be at least 600 mm SL) that was caught in about 600 feet (183 meters) of water south of Islamorada in the Florida Keys.

Material examined. Sixty-three specimens, 38 to 442 mm SL. **NORTH CAROLINA:** GMBL 78–31 (1 specimen: 360 mm SL), UF 234645 (1: 52). **SOUTH CAROLINA:** GMBL 78–13 (1: 373), GMBL 79–40 (1: 441), GMBL 79–65 (1 of 2: 349), GMBL 84–45 (1: 414), GMBL 96–7 (1: 442), UF 27790 (1: 429), UF 46523 (1: 400). **FLORIDA KEYS:** GMBL 09–011 (4: 38–58).

GULF OF MEXICO (exact locality unknown): FMNH 64440 (1: 318). **FLORIDA (GULF OF MEXICO):** FMNH 46468 (2: 161–175), FMNH 47917 (1: 165), FSBC 11807 (1: 403), FSBC 12052 (1: 61), TU 2724 (1: 164), UF 228649 (1: 155), USNM 157788 (1: 235). **ALABAMA:** USNM 134189 (holotype: 108). **LOUISIANA:** ANSP 103916 (1: 243), TCWC 7435.11 (1: 139), TU 82235 (1: 82), TU 82458 (1: 138), TU 82470 (2: 91–96), UF 38551 (2: 330–345), UF 204460 (1: 149), UF 216253 (4: 125–151), UF 228731 (2: 110–187). **TEXAS:** FMNH 47918 (1: 218), GMBL 77–277 (10: 97–285), GMBL 77–278 (1: 139), GMBL 77–292 (1: 134), TNHC uncatalogued (1: 308), TNHC uncatalogued (1: 315), TU 6800 (1: 118), UF 93567 (1: 240), UF 228728 (2: 105–154). **MEXICO (CAMPECHE BANK):** UF 28457 (1: 106), UF 28492 (1: 325), UF 28505 (1: 265). **TOBAGO:** GMBL 68–63 (1: 202). **GUYANA:** ANSP 103914 (1: 305). **SURINAME:** UF 217146 (1: 130).

Hemanthias peruanus (Steindachner, 1875)

Splittail Bass; Rose Threadfin Bass

Fig. 12; Tables 2–7; Map 5

Anthias (Hemanthias) peruanus Steindachner, 1875:378 (original description; lectotype, herein designated, MCZ 10232, 248 mm SL; type locality, herein clarified, off Paita, Peru).

Pronotogrammus peruanus (Steindachner): Jordan and Eigenmann, 1890:413 (new combination).

Hemianthias peruanus (Steindachner): Jordan and Evermann, 1896:1222 (incorrect spelling of generic name; description).

Hemanthias peruanus (Steindachner): Fitch, 1982:3, fig. 3 (species account, illustration).—Heemstra, 1995:1597 (species account).

Diagnosis. A species of *Hemanthias* distinguishable from the other members of the genus by the following combination of characters. Lower jaw usually without scales (some specimens with a few scales posteriorly). Sum of total gillrakers on first arch plus number of tubed lateral-line scales in individual specimens 81 to 89 (usually 83–88, mean 85.2). Lateral-line scales 48 to 59 (usually 50–56, mean 52.7). Urohyal without anteriorly projecting spine. Caudal fin forked with middle rays of both lobes longest in specimens greater than about 100 mm SL.

Description. Dorsal-fin rays X, 13 to 15 (usually 14). Anal-fin rays III, 7 to 9 (usually 8). Pectoral-fin rays 15 to 19 (usually 17 or 18). Gillrakers 9 or 10 + 22 to 25—total 31 to 35. Circum-caudal-peduncular scales 34 to 42 (usually 35–40). Procurrent caudal-fin rays 11 to 13 dorsally, 10 to 13 ventrally. Epineurals associated with first 11 or 12 vertebrae.

Internarial distance 6 to 12 times (usually 7–9 times) in snout length. Head length 32 to 39% SL. Snout length 6 to 10% SL. Orbit diameter 7 to 11% SL. Body depth at first dorsal spine 28 to 35% SL. Longest dorsal spine the third; third dorsal spine 17 to 44% SL in specimens 99 to 264 mm SL; third dorsal spine plus filament (where filament distinct from spine) 30 to 47% SL in specimens 95 to 236 mm SL; in some individuals third dorsal spine not distinctly separated from filament, third dorsal spine plus inseparable filament 26 to 53% SL in specimens 147 to 292 mm SL. Depressed anal-fin length 36 to 50% SL. Pelvic-fin length 29 to 49% SL. Upper caudal-fin lobe 32 to 51% SL. Lower caudal-fin lobe 33 to 51% SL. Fitch (1982) and Heemstra (1995) presented accounts of this species.

Coloration. Jordan and Evermann (1896:1223) described the coloration of *Hemanthias peruanus* as "rose-red, with small diffuse golden-brown spots on body and on soft dorsal, caudal, and anal." Walford (1974:119) wrote that "the color is rose red, with small golden brown spots on the body and on the soft dorsal, tail, and anal fins." Bussing and López (1994:96) wrote: "color red with golden spots over body and median fins. . . ." Allen and Robertson (1994:110) described the coloration of *H. peruanus* as "reddish pink with red spotting on back and upper part of head; sides pink with yellow blotches and spots; fins pink to orange-red" and presented a color illustration of the species (1994:117, pl. VII, fig. 5) which shows the head mainly rosy, body with ground color rose overlain considerably, particularly below lateral line, with bright yellow; iris of eye mostly yellow with considerable red peripherally; dorsal, anal, pelvic, and caudal fins basally with networks of rose enclosing yellow, distally bright yellow; pectoral fin pallid basally, mostly rosy distally.

Designation of lectotype. In the original description of *Anthias (Hemanthias) peruanus*, Steindachner (1875:382) wrote:

> *Fundort: Payta, Trujillo. Das Museum zu Cambridge (Mass.) besitzt ein Exemplar von erstgenannter Localität; das Wiener Museum von letzterer (durch Herrn Salmin).*

> [=Place(s) of origin: Payta, Trujillo. The museum in Cambridge (Massachusetts) possesses a specimen from the first mentioned locality; the Vienna Museum has one from the latter (via Herr Salmin)].

The second author examined the syntype (NMW 42406, 264 mm SL) housed in the museum in Vienna. The locality given for this specimen is Payta, Peru. From the data given in the original description, one would expect the locality to be Trujillo, Peru, which is at ca. 8°S, well south of Payta (= Paita), which is at ca. 5°S.

Although William N. Eschmeyer (*in litt.* to WDA, 08 December 2006) wrote that on his visit to the museum in Vienna he found evidence that one specimen of *Hemanthias peruanus* was sent to Cambridge (i.e., the MCZ), Steindachner (1875) apparently erred when he mentioned the presence of only a single specimen in the museum at Cambridge because Jordan and Eigenmann (1890:413) noted that they "examined two of Dr. Steindachner's original types (10232, M. C. Z.), from Payta, Peru. The largest of these is 15 inches [381 mm] in length [presumably total length] and is now in poor condition." Until relatively recently MCZ 10232 appeared to be missing from the collection and its status was not appreciated. As a result of a diligent search by Karsten Hartel and staff at the MCZ, the lot was found and digital photographs of both specimens and their associated labels were sent to WDA. The photograph of the specimens is clearly of a species of *Hemanthias* and subsequent examination of those specimens (MCZ 10232, 248 & ca. 275 mm standard lengths, 342+ & ca.400 mm total lengths; caudal fin damaged on shorter specimen, upper jaw damaged on larger one) shows them to be *Hemanthias peruanus*. Steindachner spent some time with Louis Agassiz at the MCZ and accompanied him on the Hassler Expedition to collect in waters off

South America (K. E. Hartel, *in litt.* to WDA, 7 December 2006). One of the labels with MCZ 10232 has "Payta, Peru. Hassler Exp." printed on it, lending further support to the contention that the specimens in this lot are indeed syntypes.

To unequivocally fix the name of the species to a zoological entity, we hereby designate as the lectotype of *Anthias* (*Hemanthias*) *peruanus* Steindachner, 1875, the smaller of the two MCZ syntypes (248 mm SL), which retains MCZ 10232 as its catalog number; by that action the other MCZ syntype (ca. 275 mm SL) becomes a paralectotype (with a new catalog number—MCZ 166564), as does the syntype in the museum in Vienna (NMW 42406, 264 mm SL).

Type locality. According to Article 76.2 of the ICZN (1999:87) the "place of origin of the lectotype becomes the type locality of the nominal species-group taxon, despite any previously published statement of the type locality." Consequently, the type locality of *Anthias* (*Hemanthias*) *peruanus* Steindachner, 1875, is off Paita, Peru.

Sexuality. Coleman (1983) presented evidence for protogynous hermaphroditism in *Hemanthias peruanus*, based on histological examination of the gonads of five specimens of this species, a female (92 mm SL), three males (145–181 mm SL), and an individual (140 mm SL) in the process of transforming from female to male.

Early life history. Beltrán-León and Ríos Herrera (2000) described and illustrated (p. 366, fig. 120) a 4.9-mm preflexion larva that they identified as *Hemanthias peruanus*. Carole C. Baldwin, who has studied the larvae of some eastern Pacific Anthiinae in considerable detail, informed (*in litt.* to WDA, 7 April 2005) that the identification is probably correct because the pigmentation of the larva illustrated is different from that of the larvae of *H. signifer* (the only other species of *Hemanthias* found in the eastern Pacific, see Baldwin, 1990:937, fig. 16). Beltrán-León and Ríos Herrera (2000) gave the numbers of dorsal- and anal-fin rays of the larva that they examined as "IX-15" and "III-9," respectively, counts that are infrequently encountered in *H. peruanus*, which almost always has counts of X, 14 and III, 8 for those fins.

Ecological note. Fitch (1982:6) mentioned that otoliths of *H. peruanus* "commonly are found in scats of sea lions, *Zalophus californianus*, that haul out on Islotes Island (north of La Paz), Gulf of California. . . ."

Distribution. We examined specimens collected off Mexico (Sonora) in the Gulf of California, off the Pacific coasts of Costa Rica, Panama, and Colombia, and off Ecuador and Peru in depths of 18 to 120/150 meters. Walford (1974:119) gave the range of the species as "Redondo (on Santa Monica Bay, California) to Chile," but we have found neither specimens nor records that would substantiate the extremities of that distribution. Rojas and Pequeño (1998a) declared the range to be Baja California (Hipólito Bank) possibly to Antofagasta, Chile (ca. 23.6°S), whereas Anderson and Baldwin (2000:379), based in part on Fitch (1982) and Grove and Lavenberg (1997), reported it to be "Hipólito Bank [27°N], Baja California Sur, in the Pacific and Cabo Lobos [ca. 30°N], Sonora, in the Gulf of California to Trujillo [8°S], Peru, and the Galápagos Islands." The record for the Galápagos is based on the report of a series of otoliths taken from the scats of a Galápagos sea lion (Grove and Lavenberg, 1997). McCosker and Rosenblatt (2010) included *H. peruanus* in their updated list of Galápagos fishes. Fitch (1982:3) gave the depth range as 10 to 117 meters.

Material examined. One hundred and fourteen specimens, 40 to 292 mm SL. **MEXICO (GULF OF CALIFORNIA; SONORA):** CAS 100282 (2 specimens, 77–95 mm SL), LACM 9543-3 (1: 40), LACM 51575-4 (1: 230), SIO 60-119-35A (7: 153–164). **COSTA RICA (PACIFIC):** GMBL 74-281 (1: 92), LACM 6624-6 (24: 87–190), LACM 6626 (1: 132). **PANAMA (PACIFIC):** LACM 49170-4 (4: 124–130), USNM 331724 (1: 90). **COLOMBIA (PACIFIC):** USNM 331721 (4: 80–146), USNM 331723 (1: 110). **ECUADOR:** SAIAB 80644 (5: 99–194), SAIAB 80645(8: 137–248), USNM uncat. (Field No. LK 66–128, 46: 121–292). **PERU:** MCZ 10232 (lectotype: 248), MCZ 166564 (paralectotype: ca. 275), NMW 42406 (paralecto-type: 264), USNM 200369 (2: 190–263), USNM 331722 (3: 100–165 mm SL).

Hemanthias signifer (Garman, 1899)

Spinythroat Jewelfish

Fig. 13; Tables 2–7; Map 5

Centristhmus signifer Garman, 1899:48, pl. 69, fig. 5 (original description; drawing of urohyal; lectotype, herein designated, MCZ 28771, 157 mm SL; type locality Pacific Ocean off Panama at 7°40'00"N, 79°17'54"W).

Hemianthias peruanus (non Steindachner), Wales, 1932:106 (misidentification, misspelling of generic name, first record for the United States—market at Redondo Beach, California).

Hemanthias delsolari Chirichigno, 1974:289, 336, fig. 558, Addenda:2 (in key and list, drawing, coast of Peru—from Máncora to Talara, note in Addenda that *H. delsolari = C. signifer*).

Hemanthias signifer (Garman): Fitch, 1982:3, fig. 4 (species account, illustration).— Heemstra, 1995:1598 (species account).

Diagnosis. A species of *Hemanthias* distinguishable from the other members of the genus by the following combination of characters. Posterior 50 to 70% of lower jaw with scales. Sum of total gillrakers on first arch plus number of tubed lateral-line scales in individual specimens 95 to 105 (usually 97–104, mean 100.6). Lateral-line scales 60 to 71 (usually 62–68, mean 65.2). Ventral border of urohyal with anteriorly project-ing spine in specimens greater than about 70 mm SL. Caudal fin lunate to forked with outermost rays of both lobes longest.

Description. Dorsal-fin rays X, 12 to 14 (usually 14). Anal-fin rays III, 7 or 8 (usually 8). Pectoral-fin rays 16 to 21 (usually 19). Gillrakers 9 to 11 + 24 to 27—total 33 to 38. Circum-caudal-peduncular scales 36 to 50 (usually 43–47). Procurrent caudal-fin rays 12 to 14 dorsally, 13 or 14 ventrally. Epineurals associated with first 11 vertebrae.

Internarial distance 8 to 16 (usually 10–14) times in snout length. Head length 34 to 41% SL. Snout length 7 to 10% SL. Orbit diameter 8 to 12% SL (8–9% SL in specimens greater than about 240 mm SL). Body depth at first dorsal spine 28 to 35% SL (28–31% SL in specimens less than about 145 mm SL, 30–35% SL in specimens greater than about 145 mm SL). Longest dorsal spine usually the third; longest dorsal spine 13 to 49% SL (20–49% SL in specimens greater than about 200 mm SL). Depressed anal-fin length 29 to 38% SL. Pelvic-fin length 23 to 37% SL (28–37% SL in specimens greater than about 150 mm SL). Upper caudal-fin lobe 28 to 35% SL. Lower caudal-fin lobe 29 to 33% SL. Fitch (1982) and Heemstra (1995) presented accounts of this species.

Coloration. Bussing and López (1994:96) wrote: "color red. . . ." Heemstra (1995:1598) described the coloration of *H. signifer* as:

> *rosado-rojizo; bordes distales de aletas dorsal, anal, y caudal amarillos; una franja amarilla desde el extremo del hocico hasta el borde preopercular; manchitas amarillas en el opérculo y en los flancos por debajo de la linea lateral* (= pink-reddish; distal edges of dorsal, anal, and caudal fins yellow; yellow stripe from tip of snout to preopercular border; small yellow spots on opercle and on sides below lateral line).

Designation of lectotype. There are five syntypes of *Centristhmus signifer*—MCZ 28771 (111 and 157 mm SL), MCZ 28772 (125 mm SL), USNM 120402 (ex MCZ 28771, 112 mm SL), USNM 153596 (ex MCZ 28772, 145 mm SL). To firmly associate the name with a specimen, we designate the larger specimen (157 mm SL) in MCZ 28771 as the lectotype of *Centristhmus signifer*. The other specimen (111 mm SL) in that lot has been assigned a new catalog number (MCZ 166565); it becomes a paralectotype, as does MCZ 28772 and each of the USNM syntypes.

Type locality. According to Article 76.2 of the ICZN (1999:87) the "place of origin of the lectotype becomes the type locality of the nominal species-group taxon. . . ." Consequently, the type locality of *Centristhmus signifer* Garman, 1899, is in the Pacific Ocean off Panama at 7°40′00″N, 79°17′54″W.

Early life history. Baldwin (1990:936; plate 1C; fig. 16) provided a description of the early developmental stages of *H. signifer* based on seven specimens, 10.0 to 31.5 mm SL, and included illustrations of four larvae (two of which were taken from Kendall, 1979, as *Pronotogrammus eos*). Watson (1996:892–893, fig. 7) furnished meristic, morphometric, and life-history data for this species and presented illustrations of three larvae.

Ecological note. Fitch (1982:6) mentioned that otoliths of *H. signifer* "commonly are found in scats of sea lions, *Zalophus californianus*, that haul out on Islotes Island (north of La Paz), Gulf of California. . . ."

Distribution. We examined specimens of *Hemanthias signifer* collected off California, Mexico (Sonora) in the Gulf of California, the Pacific coasts of Costa Rica, Panama, and Colombia, and off Ecuador and Peru in depths of 61 (61/76) to 400 (120/400) meters. Chirichigno (1974) noted its appearance (as *Hemanthias delsolari*) off northern Peru from Máncora to Talara. Fitch (1982:6) gave the range as "Playa del Rey, California (34°N), to off Paita, in northern Peru (5°S); 23 to 306 meters," and Anderson and Baldwin (2000:379) reported a similar range: "southern California to northern Peru."

Material examined. One hundred and nine specimens, 52–327 mm SL. **CALIFORNIA:** SU 24812 (1 specimen: 244 mm SL), LACM 36401–1 (1: 201), LACM 36944–1 (1: 250). **MEXICO (GULF OF CALIFORNIA, SONORA):** LACM 9543–2 (4: 52–58). **COSTA RICA (PACIFIC):** GMBL 73–189 (3: 101–137), GMBL 73–347 (20: 90–177), GMBL 73–348 (4: 107–124), GMBL 73–351 (1: 87), GMBL 74–273 (1: 113), GMBL 74–275 (2: 119–161), GMBL 74–276 (1: 124), GMBL 74–278 (1: 58), GMBL 74–280 (1: 116), GMBL 74–282 (1: 111), GMBL uncat. (1: 140). **PANAMA (PACIFIC):** GMBL uncat. (CANOPUS cruise, station 9; 1: 89), MCZ 28771 (lectotype: 157), MCZ 28772 (paralectotype: 125), MCZ 166565 (paralectotype: 111), USNM 120402 (ex MCZ 28771; paralectotype: 112), USNM 153596 (ex MCZ 28772; paralectotype: 145). **COLOMBIA (PACIFIC):** UF 230481 (1: 125), UF 230483 (1: 98), UF 234646 (9: 84–101). **ECUADOR:** CAS 57856 (2: 144–145), USNM 386085 (25: 72–119), USNM 386086 (11: 132–327), USNM 386087 (4: 155–295), USNM 386088 (3: 79–94), USNM 386089 (3: 248–270). **PERU:** ANSP 128194 (1: 302).

CHORANTHIAS, NEW GENUS

Table 1

Diagnosis. *Choranthias* is distinguishable from all other genera of Anthiinae covered herein by the following combination of characters. Scales ctenoid, with only marginal cteni, no ctenial bases in posterior fields (Fig. 1B). No secondary squamation. Maxilla with scales. Lateral-line scales 46 to 57. No fleshy papillae on orbital border. Anterior and posterior nares fairly well separated; internarial distance 3 to 6 times in snout length.

In addition, two derived characters distinguish the species of *Choranthias* from all other anthiines described herein (with the exception of the species of *Pronotogrammus*): (1) posterior border of anterior naris produced into a filament (species of *Pronotogrammus* also have an anterior narial filament—see **Remarks** below and Baldwin, 1990:941, fig. 19; 947) and (2) lateral line usually interrupted ventral to posterior part of soft dorsal fin (one species of *Pronotogrammus*, *P. multifasciatus*, occasionally with an interrupted lateral line).

Other characters that distinguish, so far as known, *Choranthias* from other anthiines covered herein (except species of *Hemanthias* and *Anatolanthias*) are (1) larvae with midlateral patch of internal pigment, (2) larvae with small knob or non-serrate ridge on supraoccipital, (3) first caudal vertebra with parapophyses (radiographs of three specimens of *C. salmopunctatus* reveal that two appear to possess parapophyses on the first caudal vertebra, but the third may lack them). *Choranthias* shares the above three traits with species of *Hemanthias* and the third trait with *Anatolanthias apiomycter* (larvae not known for *Hemanthias peruanus*, *C. salmopunctatus*, and *Anatolanthias apiomycter*—see Baldwin, 1990:934; 941, fig. 19; 946–947).

Description. With mouth closed, lower jaw exceeding upper jaw slightly. Premaxillae protrusile. No supramaxilla. Upper limb of preopercle serrate; angle with a spine or spinous process; lower limb almost smooth, except frequently with a large serra near angle. Teeth in jaws mostly small, but a few usually enlarged as canines. Vomer and palatines with small teeth; vomerine tooth patch without a posterior prolongation. Endopterygoids and tongue toothless.

Single dorsal fin (not divided to base at junction of spinous and soft-rayed portions, but fin may appear notched at junction); dorsal-fin rays usually X, 15 (rarely IX, 15 or X, 14). Anal-fin rays III, 7 to 9. Pectoral-fin rays 19 to 22. Caudal-fin deeply forked; principal rays 15 (8 + 7); branched rays 13 (7 + 6); procurrent rays 11 to 14 dorsally, 11 to 13 ventrally. Gillrakers well developed, total on first arch 32 to 39. Circumcaudal-peduncular scales 25 to 28. Vertebrae 26 (10 + 16; see Baldwin, 1990:947). Formula for configuration of supraneural bones etc. 0/0/2/1+1/1/ (Fig. 2B). Pleural ribs on vertebrae 3 through 10 (3–11 on one of 23 specimens of *C. tenuis*). Epineurals associated with first 11 or 12 vertebrae. No trisegmental pterygiophores associated with dorsal and anal fins.

Maxilla and interorbital (to varying extent) with scales (smallest specimen [UFES 0281, 34 mm SL] of *C. salmopunctatus* examined without scales in interorbital region). Dorsal and lateral aspects of snout, lachrymal region, lower jaw, gular region, branchiostegals, and branchiostegal membranes without scales. Soft dorsal and anal fins with or without scales basally; pectoral, pelvic, and caudal fins with scales basally and for varying distances onto fins. Pelvic axillary scales well developed. Modified scales (interpelvic process) overlapping pelvic-fin bases along midventral line.

Remarks. Randall (1980:107, table 2) indicated several Indo-Pacific species of *Plectranthias* that have interrupted lateral lines. According to Randall and McCosker (1992), the pored series of lateral-line scales usually terminates 1 to 4 scales anterior to the base of the caudal fin in species of the Indo-Pacific anthiine genus *Luzonichthys*. Because the species of *Choranthias*, *Luzonichthys*, and *Plectranthias* do not seem to be closely related, we hypothesize that an interrupted lateral line was derived independently in each of these three genera.

Anderson and Heemstra (1980:85) noted that *Pronotogrammus martinicensis* (cited as *Holanthias martinicensis*) "has the posterior border of the anterior nostril produced into a slender filament," and mentioned that, when reflected, this filament usually does not reach as far posteriorly as does the comparable filament in *C. tenuis*. Although acknowledging that the produced narial filament could have been derived independently in *C. tenuis* and *P. martinicensis*, Anderson and Heemstra (1980) wrote that the structure might indicate close relationship between the two species. Baldwin (1990), in her cladistic analysis of Atlantic and eastern Pacific Anthiinae, was unable to find any other characters that suggested *C. tenuis* and *P. martinicensis* are closely related. In fact, the hypothesis of a close relationship between those two species "requires at least seven hypotheses of independent acquisition or reversal" (Baldwin, 1990:948). We concur with her interpretation that the presence of produced narial filaments in *C. tenuis* and *P. martinicensis* was probably independently derived.

Baldwin (1990) regarded the presence of parapophyses on the first caudal vertebra of the species of *Choranthias* and *Hemanthias* (as restricted herein) to be a derived character among Atlantic and eastern Pacific anthiines, supporting the consideration of those taxa as a monophyletic group. Baldwin (1990:947) did not find parapophyses on the first caudal vertebra of serranine serranids, epinepheline serranids, or most of the Indo-Pacific anthiines that she examined, and indicated that additional study is "needed to determine if the closest affinities of the Indo-Pacific taxa with parapophyses on the first caudal vertebra are with" *Choranthias* and *Hemanthias* "or if this condition has evolved more than once within the subfamily." We have found parapophyses on the first caudal vertebra of the eastern Pacific *Anatolanthias apiomycter* and conjecture that its closest affinities are not with *Choranthias* and *Hemanthias*, particularly in view of the hypothesis proposed by Anderson et al. (1990) that *Anatolanthias*, *Luzonichthys*, and *Rabaulichthys* form a monophyletic group of genera.

Etymology. *Choranthias* (chora, room or space; anthias, a seafish) is from the Greek and is an allusion to the interruption of the lateral line in the species of this genus. The gender is masculine.

Type species. *Anthias tenuis* Nichols, 1920.

Key to the Species of *Choranthias*

1a. Soft rays in anal fin 7; lateral-line scales 46–51; total gillrakers on first arch 32–35 . . . . . .
. . . . *Choranthias salmopunctatus*
(central equatorial Atlantic: Saint Paul's Rocks)

1b. Soft rays in anal fin 7–9 (usually 8); lateral-line scales 51–57 (usually 52–56); total gillrakers
on first arch 34–39 (usually 35–37) . . . . *Choranthias tenuis*
(western North Atlantic)

Choranthias salmopunctatus (Lubbock and Edwards, 1981)

Salmon-spotted Jewelfish

Figure 14; Tables 2–7; Map 6

Anthias salmopunctatus Lubbock and Edwards, 1981:139, fig. 2 (original description,
illustration; holotype MZUSP 14596, 61 mm SL; type locality SW side of Bel-
monte Islet, Saint Paul's Rocks, NE of Natal, Brazil, central Atlantic Ocean).—
Luiz et al., 2007:1283, figs. 1 & 2 (rediscovery at the type locality, underwater
photographs).—Anderson, in press (species account).

Diagnosis. *Choranthias salmopunctatus* can be distinguished from the only other species in the genus *Choranthias*, *C. tenuis*, by the following combination of characters.
Soft rays in anal fin 7; lateral-line scales 46 to 51; total gillrakers on first arch 32 to 35.

Description. Dorsal-fin rays X, 15. Anal-fin rays III, 7. Pectoral-fin rays 20 to 22.
Gillrakers 9 to 11 + 23 to 25—total 32 to 35. Lateral-line scales 46 to 51 (only tubed
scales counted). Circum-caudal-peduncular scales 26 to 28.

In the original description, *Choranthias salmopunctatus* is described as having
the "lateral line briefly interrupted below base of about eleventh soft dorsal ray, with
one or two non-tubular scales separating anterior from posterior portion" (Lubbock
and Edwards, 1981:139–140). On one of the specimens that we examined (UFES
0281, 34 mm SL), the lateral line is not interrupted on the left side and has a number
of scales missing on the right, making it impossible to determine if the lateral line is
interrupted. Jean-Christophe Joyeux examined two other specimens in the ichthyological collections of the Universidade Federal do Espírito Santo and found that on
one (UFES 0554, 46 mm SL) the lateral line is interrupted on both sides, but that on
the other (UFES 0292, 47 mm SL) the lateral line is continuous on both sides
(Joyeux *in litt.* to WDA, 23 August 2007).

Posterior border of anterior naris produced into a filament which reaches or extends past posterior naris when reflected (almost reaching orbit on specimen of 34 mm
SL). Head length 30 to 32% SL. Snout length 6 to 8% SL. Orbit diameter 7 to 10% SL.
Body depth at first dorsal spine 27 to 29% SL. Longest dorsal spine (fourth to seventh
about equal in length) 10 to 13% SL. Depressed anal-fin length 29 to 32% SL. Pelvic-
fin length 24 to 30% SL. Upper caudal-fin lobe 29 to 30% SL (42% SL in specimen of
34 mm SL). Lower caudal-fin lobe 27 to ca. 31% SL (36% SL in specimen of 34 mm
SL). Ranges for the meristic and morphometric characters presented above are based
on Lubbock and Edwards (1981) and on our examination of three specimens. Lubbock

and Edwards (1981) presented a detailed description of *Choranthias salmopunctatus* (as *Anthias salmopunctatus*), and Anderson (in press) gave a short species account.

Coloration. Lubbock and Edwards (1981:141) presented the following description of coloration.

> In life, head and body of holotype light orange, becoming light brownish orange dorsally and pinkish ventrally; scattered salmon-pink spots on body; reddish pink stripe from snout tip to upper anterior margin of eye; another pink stripe from upper jaw through lower margin of eye to edge of operculum opposite upper pectoral fin base; below this a fainter pink stripe from angle of mouth to edge of operculum. Iris yellowish olive, lavender dorsally and ventrally. Pectoral and pelvic fins pink, the latter with yellow tinges; anal fin light salmon-pink with faint yellow basal stripe; dorsal fin deep salmon-pink with yellowish spines and two or three horizontal rows of faint yellow spots on soft rays; caudal fin deep salmon-pink with very faint yellow spots, brownish orange basally.
>
> In alcohol, fishes become brownish, paler ventrally, with a dark stripe from upper lip to anterior margin of eye; lower lip dark anteriorly; scattered pale spots sometimes visible on flanks; fins mostly pale or hyaline.

One paratype (BMNH 1980.1.22.2, 44 mm SL) displayed a broad dark stripe extending from anterior ends of premaxilla and dentary to middle of anterior border of orbit when examined in August 1992, almost 13 years after preservation. The smallest specimen (UFES 0281, 34 mm SL; collected in January 2006, examined in November 2007) that we have seen showed a distinctive dark stripe from tip of snout to orbit.

Luiz et al. (2007:1284–1285, figs. 1, 2) published three color photographs of *C. salmopunctatus* made underwater. The individuals in those photographs are mainly yellow to orange. One photograph (fig. 1) shows a mostly golden yellow individual with two bright yellow stripes on head, with dorsalmost stripe passing through eye, and a lavender stripe passing between the yellow stripes just below eye.

Ecology and ethology. Lubbock and Edwards (1981) described this species from four specimens (46–61 mm SL) collected at Saint Paul's Rocks and reported (p. 141) that it "was common on rock faces below 30 m depth. It was usually found in small shoals about 1 m from the substrate, and hid in crevices when frightened." Edwards and Lubbock (1983) considered *Choranthias salmopunctatus* (cited as *Anthias salmopunctatus*), along with the butterflyfish *Chaetodon* (*Prognathodes*) *obliquus*, to be characteristic of the sub-*Caulerpa* zone which harbors a community dominated by invertebrates. (In the shallower *Caulerpa* zone the green alga *Caulerpa racemosa* thickly invests the rock faces.)

In the *IUCN Red List of Threatened Species* (IUCN, 2011) this species was categorized as Vulnerable D2, i.e.,

> Population is characterised by an acute restriction in its area of occupancy (typically less than 100 km^2) or in the number of locations (typically less than five). Such a taxon would thus be prone to the effects of human activities (or stochastic events whose impact is increased by human activities) within a very short period of time in an unforeseeable future, and is thus capable of becoming Critically Endangered or even Extinct in a very short period.

On four relatively recent expeditions (in 1999, 2000, 2001) to Saint Paul's Rocks, Brazilian ichthyologists surveyed the fishes in tide pools and over reefs down to a depth of 62 meters, but they did not observe *Choranthias salmopunctatus* (Feitoza et al., 2003).

In early 2006, a group of Brazilian workers visited Saint Paul's Rocks, rediscovered *C. salmopunctatus*, and observed its behavior in depths of 35 to 55 meters. It was most abundant in 40 to 45 meters of water on nearly vertical rock drop-offs. *Choranthias salmopunctatus* swims in groups of five to ten individuals near crevices into which it retreats when threatened, each group moving into its own crevice when approached by a diver. There does not appear to be an exchange of individuals between groups or crevices, indicating that suitable refuge may be a factor determining the distribution of this species. Individuals of this species were observed to leave a crevice only upon the approach of a school of juveniles of the pomacentrid species *Chromis multilineata*. *Choranthias salmopunctatus* schools with *Chromis multilineata* and appears to use its similarity to the juveniles of that species for protection as it forages for plankton in open water (Luiz et al., 2007).

Distribution. *Choranthias salmopunctatus* is known only from Saint Paul's Rocks (00°55'N, 29°21'W), a group of islets on the Mid-Atlantic Ridge just north of the equator (ca. 970 kilometers northeast of Natal, Brazil). The type material was collected in depths of 30 to 35 meters (Lubbock and Edwards, 1981), and additional individuals were collected and observed "in depths between 33 and 55 m along a 400 m stretch of vertical cliffs on the western side of the rocks" (Luiz et al., 2007:1284).

Material examined. Three specimens, 34 to 46 mm SL. **CENTRAL ATLANTIC, SAINT PAUL'S ROCKS:** BMNH 1980.1.22.2 (paratype: 44 mm SL), SAIAB 985 (paratype: 46), UFES 0281 (1: 34 mm SL).

Choranthias tenuis (Nichols, 1920)

Threadnose Bass

Figure 15; Tables 2–7; Map 4

Anthias tenuis Nichols, 1920:60 (original description; holotype AMNH 7310, 68 mm SL; type locality off Bermuda).—Collette, 1962:441 (compiled; doubtful Bermuda endemic).—Anderson and Heemstra, 1980:74, 85 (in key; possible relationship to *Holanthias martinicensis* [= *Pronotogrammus martinicensis*]).— Bullock and Smith, 1991:15, fig. 6; pl. I, fig. B (species account, illustrations).— Smith-Vaniz, et al., 1999:204, col. pl. 11, fig. 66 (synonymy, Bermuda, distribution, illustration).—Anderson, 2003:1332 (species account).

Diagnosis. *Choranthias tenuis* can be distinguished from the only other species in the genus *Choranthias*, *C. salmopunctatus,* by the following combination of characters. Soft rays in anal fin 7 to 9 (usually 8); lateral-line scales 51 to 57 (usually 52–56); total gillrakers on first arch 34 to 39 (usually 35–37).

Description. Dorsal-fin rays X, 14 or 15 (almost always 15; one of 96 specimens with IX spines). Anal-fin rays III, 7 to 9 (almost always 8). Pectoral-fin rays 19 to 21 (20 in ca. 75% of counts). Gillrakers on first arch 9 to 11 + 24 to 28—total 34 to 39 (usually 35–37). Lateral-line scales 51 to 57 (usually 52–56, only tubed scales counted). Circum-caudal-peduncular scales 25 to 28.

Posterior border of anterior naris produced into a long slender filament, usually reaching to or very near orbit when reflected. Head length 28 to 32% SL. Snout length 5 to 7% SL. Orbit diameter 8 to 11% SL. Body depth at first dorsal spine 26 to 32% SL. Longest dorsal spine usually fourth, 11 to 13% SL. Depressed anal-fin length 30 to 38% SL. Pelvic-fin length 21 to 32% SL. Upper caudal-fin lobe 29 to 46% SL. Lower caudal-fin lobe 29 to 48% SL. Anderson (2003) presented a brief species account.

Coloration. Bullock and Smith (1991:207, pl. I, fig. B) presented a color photograph of an 83-mm specimen (FSBC 11982) of *Choranthias tenuis* collected in the eastern Gulf of Mexico and stated (p. 16): "Body rosy color with no distinct markings; iris yellow and fin membrane with a series of spots. Caudal tips and center of caudal fin scarlet." (The specimen in their photograph does not have scarlet on the tips of the caudal fin lobes nor in center of caudal fin.)

A 61-mm SL specimen (BAMZ 1990–083–121), collected off Bermuda, that was stored in alcohol in which the color preservative ionol (see Waller and Eschmeyer, 1965) had been added displayed the following coloration: pinkish red dorsally, yellow ventrally; caudal fin yellow with reddish tips; yellow-pink stripe through lower iris; mouth and snout yellow. The above description is based both on a note made by the collector, L. S. Mowbray, and on observations by PCH in June of 1973, almost three years after the date of collection.

The following description is based on WDA's examination of a color transparency provided by William F. Smith-Vaniz of a freshly collected Bermuda specimen (ANSP 133079). Snout pale pink above a broad yellow stripe that runs posteriorly from anterior end of snout to eye. Dorsoposterior part of head mostly red orange to rosy. Ventral half of head pallid except anterior end of dentary pale yellow. Iris mostly bright yellow, but with considerable dark purple on outer portions of dorsal and ventral curvatures. Body red orange dorsally, purplish-pink from about level of lateral line to mid-ventral line, except region anterior to anal fin and ventral to dorsalmost point of insertion of pectoral fin pallid. Interradial membranes of dorsal and anal fins with columns of oblong to elliptical greenish spots. Spinous dorsal fin suffused with rosy purple ventrally, green and pale purple dorsally; soft dorsal and anal fins suffused with pale purple. Membranes extending beyond tips of dorsal spines rose and purple. Pectoral and pelvic fins almost entirely pallid. Posterior one-third of upper and lower lobes of caudal fin and posterior one-half of middle caudal-fin rays blood red; most of remainder of fin dull orange basally and pale green more distally.

Early life history. Baldwin (1990:936–939; fig. 17) described the early developmental stages of *Choranthias tenuis* (cited as *Anthias tenuis*) based on 12 specimens, 3.4 mm NL to 15.0 mm SL, and included illustrations of two larvae (one of which was taken from Kendall, 1979), and Richards et al. (2006) presented an account of the early life history of the species (as *Anthias tenuis*).

Distribution. We have examined specimens collected from the Atlantic off Bermuda and North Carolina, from the Gulf of Mexico off Florida and Yucatán (Mexico), from Mona Passage off Puerto Rico, and from the southern Caribbean Sea off Venezuela and Colombia in depths of 56 to 914 meters. A single specimen of ca. 24 mm SL (USNM 246686) was taken between the surface and 100 meters off Bermuda. Most specimens were taken in less than 150 meters, but one (UF 228616, 64 mm SL) was collected in 914 meters off Venezuela (OREGON station 4416).

Bullock and Godcharles (1982) reported a male of this species (FSBC 11982, 83 mm SL) that was regurgitated by a Speckled Hind, *Epinephelus drummond-hayi*, caught by hook-and-line in the eastern Gulf of Mexico at a depth of 77 meters. William F. Smith-Vaniz informed (*in litt.* to WDA, 13 March 2000) that a good series of *tenuis* had been taken a short time before (with Sabiki Rigs) during a U.S. Geological Survey cruise on the outer continental shelf off Alabama. Weaver et al. (2006a), in a paper on results of deep reef-fish surveys made during submersible dives in the northwestern Gulf of Mexico, noted (p. 87) that they saw 2745 individuals of *tenuis* at McGrail Bank and 2300 at Alderdice Bank. Cervigón (1991:374) mentioned five specimens (73–94 mm SL) collected north of Punta Chuspa, Venezuela, in ca. 100 meters of water. Ross Robertson and Carole Baldwin sent WDA a digital photograph (Figure 15 herein) of an 85-mm-SL specimen (USNM 406397) of *Choranthias tenuis* collected off Curaçao (near 12°05.1'N, 68°53.9'W; depth 462 feet [141 meters]; 31 May 2011).

Material examined. One hundred and nine specimens, ca. 24 to 90 mm SL. **BERMUDA:** AMNH 7310 (holotype: 68 mm SL), ANSP 133079 (7: 40–56), BAMZ 1990–083–121 (1: 61), FMNH 48498 (1: 76), USNM 246686 (1: ca. 24). **NORTH CAROLINA:** FMNH 66000 (10: 55–78), FMNH 104893 (1: 65), UF 39797 (1: 50), UF 39828 (7: 28–63), UF 40771 (1: 48). **FLORIDA (GULF OF MEXICO):** FSBC 11982 (1: 83). **MEXICO (GULF OF MEXICO—OFF YUCATÁN):** UF 204122 (10: 55–64), UMMZ 174150 (22: 51–68). **PUERTO RICO (MONA PASSAGE):** GMBL 63–56 (2: 49–60). **VENEZUELA (CARIBBEAN SEA):** FMNH 70691 (3: 74–78), UF 228616 (1: 64), UF 228729 (2: 52–69), UF 229809 (2: 77–90). **COLOMBIA (CARIBBEAN SEA):** FMNH 70673 (23: 58–79), USNM 326409 (12: 63–75).

ANTHIAS BLOCH, 1792

Table 1

Anthias Bloch, 1792:97 (type species *Labrus anthias* Linnaeus, 1758:282, by absolute tautonymy).—Anderson and Baldwin, 2000:370 (synonymy, diagnosis, description, key to the species).

Aylopon Rafinesque, 1810:52 (type species *Labrus anthias* Linnaeus, 1758:282, by virtue of the fact that a replacement name retains the type of the prior name; *Anthias* incorrectly regarded as preoccupied by *Anthia* Weber, 1801, a genus of Coleoptera).

Diagnosis. *Anthias* is distinguishable from all other genera of Anthiinae covered herein by the following combination of characters. Scales ctenoid, with only marginal cteni, no ctenial bases in posterior fields (Fig. 1B). No secondary squamation. Maxilla with scales. Lateral line complete, not interrupted, tubed scales 31 to 48. Circum-caudal-peduncular scales 16 to 25. No fleshy papillae on border of orbit. Anterior and posterior nares closely set on each side of snout; posterior border of anterior naris produced into a short flap but never into a long filament. Vomerine tooth patch without a well-developed posterior prolongation.

Description. Premaxillae protrusile. No supramaxilla. Preopercle serrate but without antrorse spines on lower limb. Outer teeth in jaws mostly conical; inner teeth mostly villiform or cardiform; some enlarged as canines. Vomer and palatines with teeth; vomerine tooth patch without a well-developed posterior prolongation. Endopterygoids usually toothless. Tongue with or without teeth (present on tongue in almost all *A. menezesi*, about half of *A. cyprinoides* examined, very frequently in *A. nicholsi*, occasionally in *A. asperilinguis*, and rarely in *A. anthias*).

Single dorsal fin, not deeply notched at junction of spinous and soft-rayed portions. Dorsal-fin rays X, 13 to 16 (usually 14 or 15 soft rays, most frequently 15; very rarely with XI spines). Anal-fin rays III, 6 to 8 (7 in about 98% of specimens). Pectoral fin approximately symmetrical, with 16 to 22 rays. Caudal fin lunate to deeply forked; principal rays 15 (8 + 7); branched rays 13 (7 + 6); procurrent rays 8 to 11 dorsally, 7 to 11 ventrally. Gillrakers well developed, total on first arch 37 to 48. Vertebrae 26 (10 + 16). First caudal vertebra without parapophyses. Formula for configuration of supraneural bones, etc. 0/0/2/1+1/1/ (Fig. 2B), except *A. nicholsi* rarely with slightly different placement of the supraneural bones. Pleural ribs on vertebrae 3 through 10. Epineurals associated with first 11 to 13 vertebrae. No trisegmental pterygiophores associated with dorsal and anal fins.

Lateral line complete, extending to at least base of caudal fin, running parallel to dorsal body contour a few scale rows ventral to dorsal-fin base, then curving rather precipitously ventral to posterior end of dorsal-fin base to continue near midlateral axis of caudal peduncle. Most of head, including maxilla, dorsum of snout, and interorbital region with scales. Soft dorsal and anal fins more or less scaly at their bases; pectoral, pelvic, and caudal fins with scales basally.

Remarks. Katayama and Amaoka (1986:221) wrote that "*Anthias helenensis* and the related Atlantic species . . . seem to constitute a genus distinct from *Pseudanthias*." Other workers (e.g., Randall and Allen, 1989) have, as a consequence, considered *Anthias* as restricted to the Atlantic. We further restrict *Anthias* to exclude the congeneric species formerly known as *A. tenuis* and *A. salmopunctatus*, which we assign to our new genus, *Choranthias,* and expand it to also include *A. cyprinoides* (placed originally in *Holanthias*) and the eastern Pacific species *A. noeli.* According to Katayama and Amaoka (1986), *Anthias helenensis* and related Atlantic species of *Anthias* differ from the Pacific species formerly placed in this genus in certain internal features (perhaps most significantly in number of supraneural bones), in reaching a larger size, and in having the lateral line form an angle below the last soft dorsal ray. Although we have not found a character that is clearly synapomorphic for the eight species we include in *Anthias,* all eight are extremely similar morphologically and appear to form a natural group.

Key to the Species of *Anthias*

1a. Lateral-line scales 31–34; sum of lateral-line scales plus total gillrakers on first arch, in individual specimens, 71–76; caudal-fin lobes moderate (length of upper lobe 31–49% SL) .... *Anthias nicholsi* (western North Atlantic)

1b. Lateral-line scales 36–48; sum of lateral-line scales plus total gillrakers on first arch, in individual specimens, 75–88; caudal-fin lobes moderate to well produced (length of upper lobe 32–110% SL) 2

2a. Longest dorsal-fin spine (usually third) 13–30% SL, 18–30% SL in specimens more than ca. 100 mm SL; third dorsal-fin spine typically with well-developed filament which may be up to 19% SL; lower caudal-fin lobe usually longer than upper............... *Anthias anthias* (eastern Atlantic, including the Mediterranean and Adriatic seas)

2b. Longest dorsal-fin spine (usually fourth or fifth) 10–20% SL; fin membrane usually extending as a short filament at tip of each dorsal spine, but never produced to the extent seen in *A. anthias*; upper caudal-fin lobe usually longer than lower 3

3a. Soft dorsal-fin rays usually 14 (15 in 1 of 23 specimens); midline of gular region and lower jaw well covered with scales; pelvic-fin length 27–41% SL .... *Anthias woodsi* (western North Atlantic)

3b. Soft dorsal-fin rays usually 15 (rarely 16); gular region naked; lower jaw naked or only partly covered with scales; pelvic-fin length 33–76% SL 4

4a. Pectoral-fin rays 19–21 (usually 20); longest dorsal-fin spine 10–15% SL; upper caudal-fin lobe 32–45% SL; pelvic-fin length 33–44% SL 5

4b. Pectoral-fin rays 17–20 (usually 18 or 19); longest dorsal-fin spine 12–20% SL; upper caudal-fin lobe 39 to >70% SL; pelvic-fin length 33–76% SL 6

5a. Circum-caudal-peduncular scales 18 or 19; lower caudal-fin lobe 37–50% SL; upper caudal-fin lobe 37–45% SL; length of anal-fin base 17–19% SL; endopterygoids and tongue without teeth; posterior margin of anal fin usually rounded (occasionally more or less angulate) .... *Anthias helenensis* (eastern South Atlantic: north of the Island of Saint Helena)

5b. Circum-caudal-peduncular scales 20–24; lower caudal-fin lobe 31–35% SL; upper caudal-fin lobe 32–37% SL; length of anal-fin base 15–17% SL; endopterygoids occasionally with teeth; tongue with or without teeth; posterior margin of anal fin angulate .... *Anthias cyprinoides* (eastern South Atlantic: southwest of the Island of Pagalu)

6a. Total gillrakers on first arch 41–48; tongue usually with teeth, teeth usually in narrow elongated patch .... *Anthias menezesi* (western South Atlantic)

6b. Total gillrakers on first arch 37–41; tongue usually without teeth...................... 7

7a. Soft dorsal fin and usually soft anal fin without produced rays; depressed anal-fin length 28 to >34% SL; circum-caudal-peduncular scales 17 or 18; lateral-line scales 36–41; two of largest individuals examined (of 10 known specimens) with teeth on tongue .... *Anthias asperilinguis* (western North Atlantic)

7b. Two or more soft dorsal-fin rays and one or more soft anal-fin rays produced; depressed anal-fin length 32–43% SL; circum-caudal-peduncular scales 22–25; lateral-line scales 38–46; no teeth on tongue .... *Anthias noeli* (eastern Pacific: Ecuador, Galápagos Islands, Cocos Island)

Anthias anthias (Linnaeus, 1758)

Swallowtail Seaperch

Fig. 16; Tables 2–7; Map 7

Labrus anthias Linnaeus, 1758:282 (original description; no types known; type locality herein restricted to southern Europe, i.e., the Mediterranean Sea).

Anthias sacer Bloch, 1792:99, pl. 315 (original description; illustration; holotype ZMB 8720 [stuffed]; type locality Mediterranean Sea).

Aylopon ivicae Guichenot, 1868:81 (original description; two syntypes MNHN 4313 & 4314; type locality Mediterranean Sea, off Balearic Islands [Ibiza] and Malta).

Aylopon hispanus Guichenot, 1868:81 (original description; holotype MNHN 4315; type locality Mediterranean Sea off Spain [Algeciras Bay]).

Aylopon rissoi Guichenot, 1868:82 (original description; holotype MNHN 4216; type locality Mediterranean Sea off France [Nice]).

Aylopon nicaeensis Guichenot, 1868:83 (original description; three syntypes MNHN 0226; type locality Mediterranean Sea off France [Nice]).

Aylopon canariensis Guichenot, 1868:84 (original description; three syntypes MNHN 4316 & 4317; type locality Atlantic Ocean off Madeira and Canary Islands).

Aylopon algeriensis Guichenot, 1868:84 (original description; four syntypes MNHN 4318; type locality Mediterranean Sea off Algeria).

?*Anthias sacer* var. *brevipes* Bellotti, 1879:36 (original description; holotype destroyed in 1943 [*fide* Eschmeyer, 1998:277]; type locality off Nice, France).

Anthias mundulus Johnson, 1890:452 (original description; two syntypes BMNH 1890.5.31.2-3; type locality Madeira [Funchal Bay]).

Anthias anthias: combination used by many authors, including Anderson, in press (species account).

Diagnosis. A species of *Anthias* distinguishable from the other members of the genus by the following combination of characters. Lateral-line scales 37 to 44. Sum of lateral-line scales plus total gillrakers on first arch, in individual specimens, 77 to 86. Longest dorsal-fin spine (usually third) 13 to 30% SL, 18 to 30% SL in specimens more than ca. 100 mm SL. Third dorsal-fin spine typically with well-developed filament which may be up to 19% SL. Pelvic-fin length 34 to 74% SL (53–74% SL in specimens more than ca. 105 mm SL). Lower caudal-fin lobe usually longer than upper.

Description. Dorsal-fin rays X, 13 to 16 (usually 15). Anal-fin rays III, 7. Pectoral fin rays 18 to 22 (usually 18–20). Gillrakers 11 to 14 + 27 to 33—total 38 to 46. Lateral-line scales 37 to 44 (usually 37–41). Circum-caudal-peduncular scales 18 to 23 (usually 21 or 22).

Vomerine tooth patch roughly triangular with posterior border slightly concave to convex but without posterior prolongation. Endopterygoids toothless. Tongue rarely with teeth. Pelvic fins and caudal-fin lobes elongated. Individuals greater than ca. 80 or 90 mm SL with rays in middle of soft dorsal and middle of soft anal fins elongated. Internarial distance 8 to 17 times in snout length. Head length 30 to 37% SL. Snout length 5 to 10% SL. Orbit diameter 8 to 12% SL. Body depth at first dorsal spine 29 to 39% SL (36–39% SL in specimens more than ca. 140 mm SL). Depressed anal-fin length 32 to 51% SL (40–51% SL in specimens more than ca. 120 mm SL). Upper caudal-fin lobe 34 to >73% SL (41 to >73% SL in specimens more than ca. 100 mm SL). Lower caudal-fin lobe 33 to 75% SL. Anderson (in press) gave a short species account.

Coloration. Lowe (1843:24, pl. IV) presented a color painting and described the coloration of *Anthias anthias* (as *Anthias sacer*), writing:

> Colour fine pink or rosy, with a lilac tint; mottled along the ridge of the back with indistinct spots of dusky olive-yellow, which extend a little way down the sides, but grow paler, and presently blend into a yellow tint. Towards the belly pearly-whitish, iridescent.
>
> Sides of the head rosy, with three yellow or olive-yellow horizontal bands; one close above, another through the middle of the eye, ending between the two lower spines of the opercle; the third under the eye, and ending in a yellow spot or patch at the base of the pectoral fins. The lips are rosy. The iris chiefly pale-violet or lilac, on a silver ground. The dorsal and anal fins are rosy along their base, bordered with yellow; the *laciniae* of the former yellow. The pectoral fins are pale scarlet rather than rosy. The produced fore-part of the ventral fins is bright yellow, orange towards the tip: their spine or fore-edge pink or rosy; their hind-part white, beautifully spotted with yellow. The caudal fin is yellow, with the outer edges pink; the filaments and middle often orange.

Smith (1981) stated that *A. anthias* is "generally red, with yellow and silver marbling; belly rose; 3 yellow lines on sides of head between eye and opercle," and Tortonese (1986:781) wrote that it is "red or pink, three yellow lines on sides of head; often brownish blotches on back." Brito (1991:fig. 81) and Brito et al. (2002:229, foto

215) presented color photographs of *A. anthias*. Kuiter (2004:16) provided three color photographs—two of males and one showing two females.

The following description is based on a color transparency (received from James K. Dooley) of a specimen of *Anthias anthias* (117 mm SL) caught in 1983 in 200 meters off Gran Canaria, Canary Islands. Dorsum of head mostly yellow; lateral and ventral parts of head pink to pallid except gular region and ventralmost branchiostegals yellow (yellow extending onto chest and leading edge of pelvic fin). Three bright yellow oblique stripes on head. Ventralmost stripe beginning on upper lip passing ventral to eye to terminate on pectoral-fin near its base. Middle stripe beginning just anterior to eye, interrupted by eye, then continuing to posterior part of opercle. Dorsalmost stripe (continuous dorsally with yellow of head) beginning posterior to eye and terminating in pectoral region. Iris with pinkish purple on outer dorsal curvature, central to that pigment two concentric circles of pigment—outer blue, inner bright yellow. Body yellowish dorsally, suffused with purplish pink, midbody mostly purplish pink, ventrally body mostly pallid. Spinous dorsal fin mostly pallid, but filaments trailing from dorsal spines bright yellow. Soft dorsal fin mostly pallid basally, pale blue with dull yellow to brownish spots in middle of anterior three-fourths of fin, dull yellow distally and over most of posterior one-fourth of fin. Anal fin mostly with dull yellow to brownish spots overlying pale blue, distal portions of anterior anal soft rays and associated membranes bright yellow. Caudal fin with outermost principal dorsal and ventral rays pallid, most of fin pale yellow proximally and dull yellow distally. Pectoral fin pallid except for bright yellow on base and most proximal part of middle rays. Leading edge of pelvic fin and prolonged parts of filamentous rays bright yellow, remainder of fin with dull yellow to brownish spots.

Type locality. In the original description Linnaeus (1758:282) stated "Habitat in Europa Meridionali [= southern Europe] & America." Because *Anthias anthias* is not known from America, we restrict the type locality to southern Europe, i.e., the Mediterranean Sea.

Sexuality and sexual dimorphism. Based on histological examination of the gonads of nine specimens (eight males, 105–116 mm SL; one female, 98 mm SL), Reinboth (1964) reported that this species is protogynous. Kuiter (2004:16) stated that "Males differ from females in having long and broad ventral fins, usually ending in yellow or bright red, but variable with locations and sometimes ventral fins are all yellow."

Reproduction. Tortonese (1986:781) wrote that *A. anthias* reproduces in spring and summer (with local variations), and Bauchot (1987:1306) noted that reproduction is in spring and summer (July–September on the coast of the Maghreb).

Ecological notes. Tortonese (1986:781) gave the habitat as "rocks, gravels, submarine caves on continental shelf and upper slope to about 200 m." Dooley et al. (1985:16) stated that *A. anthias* is known from off three of the islands in the Canaries, but is "Not a common canarian inshore species." In contrast, Brito (1991:94) and Brito et al. (2002:215) reported that this species is common in the Canary Islands in depths of 30 to 300 meters over rocky bottoms and in caves in shallower water. Patzner et al. (1992:104–105) observed two groups of *Anthias anthias* in the Azores "at Monte da Guia, Faial, in front of caves at 25 and 30 m depth, respectively."

Tortonese (1986:781) stated that this species is nocturnal, feeds on crustaceans and small fishes, and may reach 270 mm SL (usually 120–180 mm SL). (The largest specimen we examined is 167 mm SL.) Bauchot (1987:1306) noted that *A. anthias* is part of the bycatch in commercial (particularly in Sicily) and sports fishing and is regularly present in the markets of Morocco—but only occasionally to rarely elsewhere.

Cape Verde Basslet. Kuiter (2004) under the heading "Cape Verde Basslet" reported a morph that closely resembles *Anthias anthias* but which may be specifically distinct from it. He provided an excellent photograph of a 27-cm (male?) specimen of this form collected off the Cape Verde Islands. Commenting on this morph, Kuiter (2004:16) wrote:

> Eastern Atlantic. A deep water species, known from few specimens from Marseille to Cape Verde. Was thought to be a variation of *Anthias anthias*. Its 3rd dorsal spine is not greatly elongated. Has large eye and caudal fin is forked, its lobes broadly rounded with filaments at tips, elongating with age. Ventral fin long and pointed. Dorsal and anal fins are pointed posteriorly. Colour yellow with pink markings. Length to about 28 cm.

Specimens of *Anthias anthias* that we have studied share some of the characteristics of the Cape Verde Basslet (elongated ventral [pelvic] fin and dorsal and anal fins pointed posteriorly (i.e., rays in middle of soft dorsal and middle of soft anal fins elongated in individuals greater than ca. 80 or 90 mm SL), but we have not examined any specimens that are clearly representative of the Cape-Verde morph. The Cape-Verde morph may be an intraspecific variant rather than a different species.

Distribution. We have examined specimens collected off Italy, the Azores, Madeira, Sierra Leone, Liberia, Ghana, Gabon, Congo Republic, and Namibia in depths of 100 to 220 (190/220) meters. The second author observed this species in 14 to 16 meters of water at Formigas, Azores, in June 1990. Literature reports give the geographic range as the Adriatic, Mediterranean, and eastern Atlantic, from Portugal and the Azores southward to northern Namibia, including off-lying islands and banks, in depths down to 320 meters (Maul, 1976; Penrith, 1978; Smith, 1981; Dooley et al., 1985; Tortonese, 1986; Maugé, 1990; Brito, 1991; Santos et al. 1997; Afonso et al., 1999; Allué et al., 2000; Bianchi et al., 2000; Brito et al., 2002; Edwards et al., 2003; Menezes et al., 2004; Anderson, in press). Maul (1976:40) wrote that "*A. anthias* is extremely common on rocky bottom between 100 and 200 m depth in the warmer parts of the eastern North Atlantic."

Material examined. One hundred and sixty-nine specimens, 31 to 167 mm SL. **EUROPE** (Bonaparte Collection): USNM 2206 (3 specimens: 79–98 mm SL). **ITALY:** SU 1975 (1: 107), USNM 48445 (1: 98). **AZORES:** USNM 94454 (2: 122–142). **MADEIRA:** AMNH 51673 (5: 118–129), SU 24027 (2: 118–121). **SIERRA LEONE:** USNM 273634 (3 of 10: 81–150), USNM 326403 (12: 81–132). **LIBERIA:** SAIAB (RUSI) 17160 (1: 115), USNM 326400 (50: 54–122), USNM 326404 (5: 75–95). **GHANA:** USNM 273630 (4: 99–111), USNM 326405 (11: 47–86). **GABON:** USNM 326401 (11: 45–68), USNM 326402 (1: 72), USNM 326406 (32: 31–158), USNM 326407 (19: 37–94). **CONGO REPUBLIC:** USNM 326408 (5: 67–141). **NAMIBIA:** IIPB 673/1981 (1: 167).

Anthias asperilinguis Günther, 1859

Jeweled Gemfish

Fig. 17; Tables 2–7; Map 6

Anthias asperilinguis Günther, 1859:89 (original description; holotype BMNH 1974.10.4.1, 143 mm SL; type locality South America [Atlantic coast]). —Anderson and Heemstra, 1980:79 (species account; illustration).—Anderson, 2003:1330 (species account).

Diagnosis. A species of *Anthias* distinguishable from other members of the genus by the following combination of characters. Soft dorsal-fin rays 15. Pectoral-fin rays 18 or 19. Lateral-line scales 36 to 41. Total number of gillrakers on first arch 38 to 40. Sum of lateral-line scales plus total number of gillrakers on first arch, in individual specimens, 75 to 79. Circum-caudal-peduncular scales 17 or 18. Longest dorsal-fin spine (usually fourth or fifth) 12 to 16% SL; fin membrane extending into a short filament at tip of each dorsal spine, but filaments never greatly produced. Soft dorsal fin and usually soft anal fin without produced rays. Depressed anal-fin length 28 to >34% SL. Pelvic-fin length 35 to >64% SL. Anterior part of lower jaw and gular region naked. Tongue usually without teeth.

Description. Dorsal-fin rays X, 15. Anal-fin rays III, 7. Pectoral-fin rays 18 or 19 (usually 19). Gillrakers 11 to 13 + 26 to 28—total 38 to 40. Lateral-line scales 36 to 41 (usually 37 or 38).

Vomerine tooth patch roughly triangular with posterior border frequently convex (patch more or less chevron-shaped in specimen of 54 mm SL). Endopterygoids and tongue usually toothless. Internarial distance 10 to 15 times in snout length. Head length 34 to 37% SL. Snout length 8 to 10% SL. Orbit diameter 11 to 14% SL. Body depth at first dorsal spine 36 to 41% SL. Upper caudal-fin lobe damaged on almost all specimens examined (in a specimen of 54 mm SL, ca. 49% SL). Lower caudal-fin lobe 41 to >46% SL. Anderson and Heemstra (1980) gave a detailed description of *A. asperilinguis*, and Anderson (2003) presented a short species account.

Coloration. Günther (1859:90–91) gave the following description of the coloration of the holotype:

> The specimen is rather discoloured; but there is every probability of its having been rose-coloured, with parallel oblique and longitudinal shining golden bands following the series of scales. There are visible on the sides of the head two oblique bands of indistinct colour, one running from the back edge of the orbit across the operculum, the other from below the eye to the inferior part of the operculum.

Boulenger (1895:326) described the coloration as "Red, with golden stripes along the series of scales" and included a black and white illustration (pl. XIII) of the holotype.

Sexuality. William A. Roumillat examined histological sections of two specimens of *A. asperilinguis* and found one to be an immature female (GMBL 63-55: 142 mm SL) and the other to be a mature spawning female (GMBL 63-55: 150 mm SL). Unfortunately, study of those sections does not help determine whether this species is hermaphroditic or gonochoristic.

Distribution. *Anthias asperilinguis* has been collected off Venezuela, the Guianas, and northeastern Brazil (just north of the equator) in depths of 229 to 320 meters.

Material examined. Known only from the ten specimens (54-156 mm SL) listed by Anderson and Heemstra (1980:80). One of the lots, TABL 107587, reported by Anderson and Heemstra (1980) has been recatalogued as GMBL 63-55.

Anthias cyprinoides (Katayama and Amaoka, 1986)

Gemmed Jewelfish

Fig. 18; Tables 2–7; Map 7

Holanthias cyprinoides Katayama and Amaoka, 1986:213, fig. 1 (original description, illustration; holotype HUMZ 100019, 193 mm SL; type locality eastern South Atlantic, southwest of the Island of Pagalu [Annobón] at 03°01'S, 00°46'E).

Anthias cyprinoides: Anderson, in press (species account).

Diagnosis. A species of *Anthias* distinguishable from the other members of the genus by the following combination of characters. Soft dorsal-fin rays 15. Pectoral-fin rays 19 to 21. Lateral-line scales 38 to 43. Sum of lateral-line scales plus total number of gillrakers on first arch, in individual specimens, 77–82. Circum-caudal-peduncular scales 20 to 24. Longest dorsal-fin spine (third, fourth, fifth, or sixth—most frequently the fourth) 11 to 15% SL; fin membrane extending into a short filament posterior to tip of each dorsal-fin spine, but filaments never greatly produced. Posterior margin of anal fin angulate. Length of anal-fin base 15 to 17% SL. Pelvic-fin length 33 to 44% SL. Upper caudal-fin lobe 32 to 37% SL. Lower caudal-fin lobe 31 to 35% SL. Lower jaw and gular region naked. Endopterygoids usually without teeth. Tongue with or without teeth.

Description. Some of our meristic data differ slightly from those of Katayama and Amaoka (1986) in the original description. Where different, the data in the original description follow ours in parentheses. Dorsal-fin rays X, 15. Anal-fin rays III, 7 (III, 7 or 8, usually 7). Pectoral-fin rays 19 or 20, usually 20 (20 or 21). Gillrakers 11 to 13 + 27 to 29—total 38 to 41 (12–14 + 26–29—total 39–43). Lateral-line scales 38 to 43, rarely 43 (38–42). Circum-caudal-peduncular scales 22 to 24 (20–22).

In 11 specimens examined by us: vomerine teeth in triangular patch or in diamond-shaped patch (in one specimen) without a well-developed backward prolongation. Endopterygoids usually without teeth (teeth present in two specimens). Tongue with small teeth (in one patch in four specimens, two patches in one) or without teeth (six specimens). Internarial distance 6 to 14 times in snout length. Head length 33 to 37% SL. Snout length 7 to 9% SL. Orbit diameter 8 to 11% SL. Body depth at first dorsal spine 34 to 39% SL. Depressed anal-fin length 31 to 36% SL. Katayama and Amaoka (1986) presented a detailed description of *A. cyprinoides*, and Anderson (in press) presented a short species account.

Coloration. Katayama and Amaoka (1986:214) described the coloration from an Ektachrome transparency as: "body yellowish brown; a yellow oblique band from tip of snout passing below eye and extending to base of pectoral fin; another longitudinal yellow band runs from hind edge of orbit to opercular margin; fins yellowish grey."

Remarks. Katayama and Amaoka (1986) relegated this species to *Holanthias*, presumably on the basis of its vomerine and lingual dentition. They stated (p. 214) "prevomerine tooth patch with a well-developed backward prolongation" and "large oval patch of teeth on tongue." None of the 11 specimens that we examined has a well-developed backward prolongation of the vomerine tooth patch, and only five of the 11 have teeth on the tongue. At our request, Kunio Amaoka examined the holotype (HUMZ 100019) and one of the paratypes (HUMZ 100136) and found that in both specimens the vomerine tooth patch is approximately diamond shaped without a posterior prolongation and that the patch of teeth on the tongue is oval (Amaoka, *in litt.* to WDA, 22 November 2004). The vomerine and lingual dentition resembles that of *Anthias* rather than that of *Holanthias fronticinctus* (the type species of *Holanthias*), which has a well-developed posterior prolongation of the vomerine tooth patch and a large oval patch of teeth on the tongue. In addition, *H. fronticinctus* usually has a well-developed patch of teeth on each endopterygoid, whereas species of *Anthias* usually have edentulous endopterygoids; *cyprinoides* usually lacks endopterygoid teeth (two of 11 specimens examined with endopterygoid teeth—present on both sides in one specimen, on right side only in the other). In two other characters, absence of accessory scales and the possession of a deeply forked caudal fin, this species is also like *Anthias* rather than *H. fronticinctus*, which has small accessory scales at bases of large scales on head and body and has a convex or doubly truncate caudal fin with middle rays longest, sometimes appreciably elongated. Anderson and Baldwin (2000) reevaluated the status of *cyprinoides* and considered it to be more appropriately placed in *Anthias*. In view of the preceding, we follow Anderson and Baldwin (2000) and consider *cyprinoides* to be a species of *Anthias*.

Distribution. *Anthias cyprinoides* is known only from a few collections made southwest of the Island of Pagalu (Annobón) in the eastern South Atlantic in depths of 260 (260/261) to 589 (260/589) meters. (All of the specimens examined were collected at latitude 03°01'S; longitudes of capture were 00°44', 00°46', and 00°47'E.)

Material examined. Eleven specimens, 127 to 222 mm SL. **SOUTHWEST OF THE ISLAND OF PAGALU (ANNOBÓN):** HUMZ 100018 (paratype: 222 mm SL), HUMZ 100020 (paratype: 166), HUMZ 100137 (paratype: 215), IIPB 346–350/1999 (5: 127–194), IIPB 421/1999 (1: 162), IIPB 48–49/2000 (2: 178–222).

Anthias helenensis Katayama and Amaoka, 1986

Central Atlantic Seaperch; Rosy Gemfish

Fig. 19; Tables 2–7; Map 7

Anthias helenensis Katayama and Amaoka, 1986:215, fig. 2 (original description, illustration; holotype HUMZ 100156, 163 mm SL; type locality eastern South Atlantic north of the Island of St. Helena at 11°37'S, 05°13'W).—Anderson, in press (species account).

Diagnosis. A species of *Anthias* distinguishable from the other members of the genus by the following combination of characters. Soft dorsal-fin rays 15. Pectoral-fin rays 19 to 21. Lateral-line scales 37 to 42. Sum of lateral-line scales plus total number of gillrakers on first arch, in individual specimens, 78 to 84. Circum-caudal-peduncular scales 18 or 19.

Longest dorsal-fin spine (fourth or fifth) 10 to 13% SL; fin membrane extending into a short filament posterior to tip of each dorsal-fin spine, but filaments never greatly produced. Posterior margin of anal fin usually rounded (occasionally more or less angulate). Length of anal-fin base 17 to 19% SL. Pelvic-fin length 35 to 39% SL. Upper caudal-fin lobe ca. 37 to ca. 45% SL. Lower caudal-fin lobe ca. 37 to 50% SL. Lower jaw and gular region naked. Endopterygoids and tongue without teeth.

Description. Some of our meristic data differ slightly from those of Katayama and Amaoka (1986) in the original description. Where different, the data in the original description follow ours in parentheses. Dorsal-fin rays X, 15. Anal-fin rays III, 7. Pectoral-fin rays 19 or 20 (19–21). Gillrakers 12 + 29 to 31—total 41 to 43 (11 or 12 + 29–31—total 40–43). Lateral-line scales 37 to 41 (38–42). Circum-caudal-peduncular scales 18 or 19 (18).

Vomerine tooth patch approximately triangular, without posterior prolongation. In the four specimens examined by us: Internarial distance 6 to 12 times in snout length. Head length 34 to 37% SL. Snout length 6 to 8% SL. Orbit diameter 10 to 11% SL. Body depth at first dorsal spine 35 to 37% SL. Depressed anal-fin length 30 to 35% SL. Katayama and Amaoka (1986) presented a detailed description of *A. helenensis*, and Anderson (in press) presented a short species account.

Coloration. Katayama and Amaoka (1986:217) described the coloration of a specimen from an Ektachrome transparency as: "body chocolate-colored; each scale on side of body with a vertically elongate white spot; all fins except for pelvic fin chocolate-colored; pelvic fin pale pink." Kunio Amaoka sent PCH Ektachrome transparencies of the holotype (HUMZ 100156, 163 mm SL) and one of the paratypes (HUMZ 100157, 165 mm SL). These transparencies show the body and fins to be mostly rose colored except where blanched, not chocolate colored as stated in the original description, but as noted by Katayama and Amaoka (1986:217) a "vertically elongate white spot" is present on each scale on side of body.

Distribution. *Anthias helenensis* is known from only two localities (11°37'S, 05°13'W and 11°37'S, 05°14'W) well north of the Island of Saint Helena in the eastern South Atlantic in depths of 156 (156/165) to 460 (163/460) meters.

Material examined. Four specimens, 147 to 177 mm SL. **NORTH OF THE ISLAND OF SAINT HELENA:** HUMZ 100160 (paratype: 177 mm SL), HUMZ 100161 (paratype: 149), IIPB 418/1999 (1: 171), IIPB 419/1999 (1: 147).

Anthias menezesi Anderson and Heemstra, 1980

Spectacular Seaperch

Fig. 20; Tables 2–7; Map 6

Anthias menezesi Anderson and Heemstra, 1980:80, figs. 4, 5 (original description, illustrations; holotype MZUSP 11765, 132 mm SL; type locality off southern Brazil at 33°38'S, 51°04'W).

Odontanthias aspergilingua (non Günther, 1859): Silva, 1936:188 (*lapsus calami* for *asperilinguis*; description and illustration of specimen collected from the Abrolhos—a group of rocky shoals, at ca. lat. 18°S, SE of Caravelas, Bahia, Brazil).

Diagnosis. A species of *Anthias* distinguishable from the other members of the genus by the following combination of characters. Soft dorsal-fin rays 15 or 16. Pectoral-fin rays 17 to 19. Total number of gillrakers on first arch 41 to 48. Lateral-line scales 36 to 40. Sum of lateral-line scales plus total number of gillrakers on first arch, in individual specimens, 80 to 87. Circum-caudal-peduncular scales 16 to 18. Longest dorsal-fin spine (usually fifth or tenth) 12 to 15% SL; fin membrane extending into a short filament at tip of each dorsal spine, but filaments never greatly produced. Pelvic-fin length ca. 52 to ca. 76% SL. Upper caudal-fin lobe ca. 53 to >70% SL. Anterior and ventromedial parts of lower jaw and gular region without scales. Tongue usually with teeth.

Description. Dorsal-fin rays X, 15 or 16 (usually 15). Anal-fin rays III, 7. Pectoral fin rays 17 to 19 (most frequently 18). Gillrakers 12 to 15 + 29 to 33—total 41 to 48. Lateral-line scales 36 to 40 (most frequently 37). Circum-caudal-peduncular scales 16 to 18 (most frequently 18).

Vomerine tooth patch variable in shape—from subtriangular to almost diamond shaped—without posterior prolongation. Endopterygoids usually without teeth (teeth present in 4 of 14 specimens). Tongue usually with teeth (teeth present in 13 of 14 specimens); teeth on tongue usually in a narrow elongated patch. Pelvic fins and caudal-fin lobes elongated in larger individuals. Head length 34 to 38% SL. Snout length 7 to 10% SL. Orbit diameter 10 to 16% SL. Body depth at first dorsal spine 37 to 41% SL. Depressed anal-fin length 31 to 37% SL. Upper caudal-fin lobe ca. 53 to >70% SL. Lower caudal-fin lobe >37 to >59% SL. Anderson and Heemstra (1980) gave a detailed description of *Anthias menezesi*.

Coloration. The following description, quoted from Anderson and Heemstra 1980:81), is based on coloration of the holotype and one of the paratypes.

> Head and body pink violet above and on sides, becoming paler below. Lower part of head and ventral part of thoracic region anterior to pelvic fin yellow. Anterior tip of lower jaw dark pink. Iris striking: bright blue on dorsal and ventral (or anteroventral) curvatures; reddish on anterior (or anterodorsal) and posterior (or posteroventral) curvatures; yellow in a narrow zone anteriorly (adjacent to pupil and inside of and adjacent to reddish curvature) and in a broader zone posteriorly, or posterodorsally (adjacent to pupil and inside of and adjacent to blue and reddish curvatures). Three horizontal yellow stripes on head and anterior part of body. Ventralmost stripe beginning at anterior tip of premaxilla and extending to origin of pectoral fin, represented by yellow spots from that point posteriorly to base of caudal fin. Median yellow stripe originating at eye and extending to dorsal part of pectoral-fin base, from that point posteriorly represented by yellow spots— spots less conspicuous towards posterior part of body. Dorsalmost stripe represented by three closely aligned, horizontally oriented blotches—first of which originating above posterior half of eye and about equal to orbital diameter in length. Dorsal fin yellow with distal ends of filaments associated with spines and distal tips of soft rays pink, giving fin a thin pink border. Anal fin yellow. Pectoral fin uniform light pink. Pelvic fin yellow with basal portion pink. Caudal fin yellow with light pink distal margin, tips of caudal-fin lobes dark pink.

Remarks. Silva (1936) described and illustrated an anthiine of 18.5 cm (total?) length caught in the Abrolhos (a group of rocky shoals, at ca. lat. 18°S, SE of Caravelas, Bahia, Brazil) that he called *Odontanthias aspergilingua*. Based on Silva's description and illustration, Anderson and Heemstra (1980) concluded that he had a specimen of *Anthias menezesi*.

Distribution. *Anthias menezesi* has been collected off northeastern (lat. 01.5°S) and southern Brazil and off Uruguay (to as far south as lat. 34.2°S) in depths of 160 to 260 meters.

Material examined. The only specimens examined are the 14 (85–167 mm SL) listed by Anderson and Heemstra (1980:82).

Anthias nicholsi Firth, 1933

Yellowfin Bass

Fig. 21; Tables 2–7; Map 8

Anthias nicholsi Firth, 1933:158, fig. (original description, illustration; holotype USNM 92936, 143 mm SL, type locality off Chesapeake Light Vessel, Virginia). —Anderson, 2003:1331 (species account).

Diagnosis. A species of *Anthias* distinguishable from the other members of the genus by the following combination of characters. Lateral-line scales 31–34. Sum of lateral-line scales plus total gillrakers on first arch, in individual specimens, 71–76. Caudal-fin lobes moderate (length of upper lobe 31–49% SL).

Description. Dorsal-fin rays X, 14 or 15 (usually with 15 soft rays, rarely with XI spines). Anal-fin rays III, 6 to 8 (7 in ca. 95% of specimens). Pectoral fin rays 18 to 21 (most frequently 19). Gillrakers 11 to 13 + 27 to 31—total 39 to 44 (usually 40–43). Lateral-line scales 31 to 34 (most frequently 32). Circum-caudal-peduncular scales 17 or 18 (most frequently 18).

Internarial distance 9 to 11 times in snout length. Head length 34 to 39% SL. Snout length 6 to 11% SL. Orbit diameter 9 to 17% SL (9–13% SL in specimens greater than 70 mm SL). Body depth at first dorsal spine 34 to 44% SL. Longest dorsal spine 11 to 16% SL in specimens greater than 70 mm SL. Depressed anal-fin length 32 to 40% SL. Pelvic-fin length 33 to >46% SL (38 to >46% SL in specimens greater than 130 mm SL). Upper caudal-fin lobe 31 to 49% SL (31–39% SL in specimens greater than 100 mm SL). Lower caudal-fin lobe 29 to 46% SL (29–38% SL in specimens greater than ca. 100 mm SL). Springer and Johnson (2004:140–141, pl. 111) described and illustrated the dorsal gill arch musculature of *Epinephelus merra* and noted that the musculature of an *Anthias nicholsi* of 110 mm (USNM 151904) is very similar. Anderson (2003) presented a brief species account.

Coloration. Firth (1933:159) described the coloration of *A. nicholsi* as follows:

> In formalin the specimen is pale in color with pink shades on body and fins. The end of the lower jaw is strongly pink. The margin of the dorsal and most of the anal are bright yellow. Outer margin of ventral pink; center of fin

bright yellow; inner angle pale. A yellow stripe runs backward below the eye to the base of the pectoral, and another extends from the back of the eye to the margin of the opercle. There is an olive blotch in the middle of the back at the base of the first dorsal spine. When fresh this spot was deep blue; the head showed more or less radiating stripes of yellow and red; and caudal was whitish, margined all around with red; the body had 3 or 4 more or less well defined lengthwise yellow stripes, and was silvery white below; the body and fins were otherwise marked with pink and yellow.

Fowler (1937:300, fig. 4) depicted a specimen of *A. nicholsi* and provided a description of its coloration:

Color of back, when fresh, geranium pink to peach-blossom pink, fading paler on lower sides and under surfaces. Iris crimson to lake red, with lemon yellow ring around pupil. On back, along bases of dorsals, ill-defined band of gallstone yellow. Obscurely from upper hind part of eye poorly defined band back to suprascapula and below lateral line, where broadening out to caudal peduncle. Deep lemon yellow band from lower hind eye edge to costal region, and another includes all of preorbital back across cheek to upper prepectoral region. On tail most of scales with lemon yellow spots, producing obscurely mottled appearance. Lower and under sides of body whitish, with median lemon yellow band from symphysis to ventrals. Dorsals brilliant rose red or lake red, with upper part of fin chrome to lemon yellow, and bright gamboge blotch on each spine basally. Soft dorsal more broadly chrome to lemon yellow, this portion of rays exceptionally brilliant. Anal like soft dorsal, basally geranium pink and nearly outer ¾ of fin very brilliant lemon yellow. Pectoral salmon color. Ventral with narrow front edge rose pink, then broadly brilliant chrome yellow, and last or two innermost rays pale pink. Caudal largely brilliant chrome yellow, outer half of each lobe pink.

Bullock and Smith (1991:207, pl. I, fig. A) presented a color photograph of *A. nicholsi* and wrote (p. 14):

Body color reddish lavender; yellow markings as follows: fins; various patches on body; two stripes, one below eye from tip of snout to pectoral base, the second through eye to opercular margin. Mature fish with elongated filament [yellow in their illustration] from interspinous membrane of third dorsal spine; a short filament (lavender) behind other dorsal spines. . . .

The following description is based on examination of several color transparencies (received from George Burgess, Donald D. Flescher, Churchill Grimes, and Charles A. Wenner) of specimens of *Anthias nicholsi*. Dorsum of head mainly rosy, suffused with yellow; lateral aspect of head paler. Ventrum of head with bright yellow stripe beginning at anterior part of gular, continuing posteriorly over isthmus and chest to fuse or almost fuse with bright yellow of pelvic fins. Two bright yellow stripes on side of head; ventral one beginning anteriorly on upper lip, coursing somewhat obliquely ventral to eye to posterior end of head where it continues (sometimes broken) onto the base of the pectoral fin; dorsal stripe extending from snout (broken by eye) to posterior end of head, becoming continuous with short stripe at anterior end of body. A third yellow stripe, much less distinct, sometimes present, extending from posterodorsal aspect of orbit beyond head to terminate on body. Iris brightly colored; pigment arranged in irregular concentric circles or hemispheres; outermost pigment usually bright blue (sometimes bordered peripherally, in places, by rosy purple); innermost pigment bright yellow, sometimes overlain in small sections by rose, purplish rose, or orange. Body rosy dorsally, suffused with yellow or dull brownish yellow; paler ventrally with yellow sometimes more prominent. Fins, except pectoral, mostly yellow; dorsal and caudal fins often with an admixture of rose; only leading half of pelvic fin pigmented, remainder pallid; pectoral fin pigmented proximally adjacent to yellow stripe which terminates on pectoral-fin base, rest of fin pallid.

Sexuality. In a sample of 18 specimens, Bullock and Smith (1991) found females ranging from 71 to 139 mm SL and males from 106 to 149 mm SL. The difference in size ranges for the sexes, along with their finding an individual of 89 mm SL with both female and male tissues, strongly suggests that *Anthias nicholsi* is a protogynous hermaphrodite. Anderson and Baldwin (2000) reported similar results based on examination of histological sections of gonads of 20 specimens of *A. nicholsi* (12 females, 52.0–125 mm SL; one individual, 73.0 mm SL, transforming from female to male; and seven secondary males, 99.9–134 mm SL) and concluded on the basis of available evidence that the species is protogynous.

Reproduction. Bullock and Smith (1991), although lacking sufficient material to determine the spawning season for *Anthias nicholsi*, did find a well-developed ovary in a female captured off the Florida Keys in February, ripe females in collections made in April in the eastern Gulf of Mexico, and spent females in collections of April and June.

Early life history. Baldwin (1990:927–929; figs. 9, 10) provided a description of the early developmental stages of *A. nicholsi* based on 86 specimens, 2.0 mm NL to 24.0 mm SL, and included illustrations of four larvae (two of which were taken from Kendall, 1979, as *Anthias*—type 1), and Richards et al. (2006) presented an account of the early life history of the species.

Ecological notes. *Anthias nicholsi* has been observed to hover individually over clumps of *Oculina* and over larger boulders on the northern rim of the DeSoto Canyon in the northern Gulf of Mexico (Shipp and Hopkins, 1978). Williams and Shipp (1980) mentioned frequent sightings along the rim of the DeSoto Canyon of *A. nicholsi* in and around large schools of *Hemanthias* sp. Such groups were never seen "more than 2 m above the limestone boulders and usually were in, or just above, large crevices between the boulders" (Williams and Shipp, 1980:19). Interestingly, a 59.2-mm specimen collected off North Carolina "was removed from a crevice in a large chunk of marl" (Burgess et al., 1979:80). Perhaps, as suggested by Williams and Shipp (1980), this species enters crevices to avoid large predators. Support for this conjecture is provided by specimens of *A. nicholsi* found in the stomach contents and spewings of groupers (Bullock and Smith, 1991) and in the stomach of a Greater Amberjack, *Seriola dumerili,* (Barans et al., 1986).

Barans et al. (1986:91, 92) reported on observations made in August 1982 on *Anthias nicholsi* from a submersible that operated at about 148 kilometers due east of Charleston, South Carolina, in 188 to 207 meters of water. They noted that this species generally avoided the submersible, especially its artificial lights, by hiding in burrows and among the rocks; occasionally, however, "individuals . . . did not hide (or only slowly proceeded to a hiding place) allowing photography under full light" (p. 92). In 1982/1983 in the same general locality in 185 to 220 meters, Gutherz et al. (1995: table 2) observed 11,798 individuals of *A. nicholsi* in point (1057) and transect (10,741) counts made from a submersible, representing 36% of the total number of fishes seen in point counts and 60% of those noted in transects.

On 5 September 2002 at a locality (38°28'N, 73°31'W) off the northeastern United States, 1135 specimens of this species were collected, indicating that *A. nicholsi* is sometimes quite abundant (Moore et al. 2003:225).

Bullock and Smith (1991) found the remains of pteropods and copepods in the stomachs of 18 specimens of *A. nicholsi*.

Distribution. Anderson and Heemstra (1980:74) reported the range of *Anthias nicholsi* as "New Jersey to the Straits of Florida, in the Gulf of Mexico and from Guyana to northeastern Brazil (Pará)." During the current study we examined a single specimen from the Caribbean Sea collected off Nicaragua and two specimens caught off Nova Scotia. Specimens we examined came from depths of 73 (73/91) to 256 meters.

Markle et al. (1980) and Scott and Scott (1988) reported larvae of *Anthias* collected off southeastern Nova Scotia; those specimens were later identified as *A. nicholsi* by Baldwin (1990). Gilhen and McAllister (1981) described a 125-mm SL specimen of *A. nicholsi* caught in 190 meters off southeastern Nova Scotia on 3 July 1980. Through the courtesy of John Gilhen we have examined two other specimens of *A. nicholsi* caught off southeastern Nova Scotia in May 1986. One, 137 mm SL, was obtained on the 25th in 125 meters (surface temperature: 11°C); the other, 130 mm SL, on the 28th in 155 to 170 meters. The capture of specimens as large as these in May suggests that *A. nicholsi* is capable of overwintering at latitudes at least as far north as Nova Scotia.

Moore et al. (2003:225) reported *A. nicholsi* from four localities off the northeastern United States (the most northerly being 39°57'N, 67°30'W). Fowler (1937) reported a specimen collected about 70 miles southeast of Cape May, New Jersey. Burgess et al. (1979:80) wrote that "*A. nicholsi* is not uncommon in deeper (73–188 m) North Carolina waters." Bullock and Smith (1991) examined 18 specimens (71–149 mm SL) of *A. nicholsi* from the Florida Keys and the eastern Gulf of Mexico from depths of 146 to 427 (366/427) meters; five of those (107–143 mm SL) were regurgitated by groupers. Williams and Shipp (1980) reported this species from depths of 54 (based on observations) and 162 meters in the northeastern and eastern Gulf of Mexico.

Material examined. One hundred and nineteen specimens, 29 to 190 mm SL. **WESTERN ATLANTIC (no other data):** FMNH 70708 (2 specimens: 151–171 mm SL). **NOVA SCOTIA:** NSMC 986-150 (1: 137), NSMC 986-151 (1: 130). **NEW JERSEY:** ANSP 120369 (1: 106). **DELAWARE:** MCZ 162088 (2: 147–157). **VIRGINIA:** AMNH 12337 (paratype: 136), AMNH 20933 (paratype: 114), AMNH 77291 (3: 74–109), GMBL 57-48 (1: 190), GMBL 81-146 (4: 66–109), USNM 92936 (holotype: 143). **NORTH CAROLINA:** GMBL 59-32 (10: 52–100), GMBL 60-12 (1: 132), GMBL 81-63 (1: 127), UF 24523 (1: 60), UF 32868 (1: 140), UF 39771 (2: 102–156), UF 41867 (1: 103), UF 230345 (1: 76). **SOUTH CAROLINA:** AMNH 77310 (5: 140–162), GMBL 60-20 (1: 127), GMBL 73-304 (5: 92–126), GMBL 75-104 (1: 135), GMBL 80-01 (3: 84–135), GMBL 82-29 (13: 87–148), SHML 171 (1: 100), UF 24524 (19: 105–168), UF 230477 (1: 175). **FLORIDA (ATLANTIC):** UF 216781 (1: 101). **STRAITS OF FLORIDA:** ANSP 120368 (3: 69–147), FMNH 70693 (1: 64),

UF 79754 (2: 67–69), UF 96279 (1: 53), UF 228266 (1: 90). **FLORIDA (GULF OF MEXICO):** FSBC 11643 (1: 114), UAIC 8036.04 (15: 50–117), UF 72313 (2: 85–90). **NICARAGUA (CARIBBEAN SEA):** UF 212272 (1: 73). **GUYANA:** GMBL 68-61 (1: 122). **SURINAME:** FMNH 70707 (3: 94–155). **FRENCH GUIANA:** UF 230274 (1: 29). **BRAZIL (NORTHEAST COAST):** USNM 216269 (1: 91).

Anthias noeli Anderson and Baldwin, 2000

Rosy Jewelfish

Fig. 22; Tables 2–7; Map 3

Anthias noeli Anderson and Baldwin, 2000:372, figs. 1–5 (original description, illustrations; holotype USNM 353113, 167 mm SL; type locality seamount SE of Isla San Cristobal, Galápagos Islands).—Béarez and Jiménez Prado, 2003:109, fig. 5, table 1 (off Santa Rosa, Ecuador; color photograph; meristic and morphometric data).

Diagnosis. A species of *Anthias* distinguishable from the other members of the genus by the following combination of characters. Soft dorsal-fin rays 15 or 16. Pectoral-fin rays 18 to 20. Total gillrakers on first arch 37 to 41. Lateral-line scales 38 to 46. Sum of lateral-line scales plus total gillrakers on first arch, in individual specimens, 78 to 85. Circum-caudal-peduncular scales 22 to 25. Longest dorsal-fin spine (fourth or fifth) 14 to 20% SL; fin membrane extending into a short filament at tip of each dorsal spine, but filaments never greatly produced. Two or more soft dorsal-fin rays and one or more soft-anal fin rays produced. Depressed anal-fin length 32 to 43% SL. Pelvic-fin length 33 to 55% SL. Upper caudal-fin lobe longer than lower. Upper caudal-fin lobe 39 to 60% SL. Lower jaw usually naked, but some specimens with a few scales posteriorly. Gular region naked. Tongue without teeth.

Description. Dorsal-fin rays X, 15 or 16 (usually 15). Anal-fin rays III, 6 or 7 (usually 7). Pectoral-fin rays 18 to 20 (usually 19). Gillrakers 10 to 12 + 27 to 30—total 37 to 41. Lateral-line scales 38 to 46 (usually 39 to 44). Circum-caudal-peduncular scales 22 to 25 (usually 23 or 24).

Vomerine tooth patch chevron-shaped to roughly triangular without a posterior prolongation (teeth extremely small on specimens less than ca. 110 mm SL). Endopterygoids without teeth. Internarial distance 6 to 12 (usually 7–10) times in snout length. Head length 37 to 43% SL. Snout length 6 to 9% SL. Orbit diameter 11 to 15% SL. Body depth at first dorsal spine 35 to 42% SL. Upper caudal-fin lobe 39 to 60% SL. Lower caudal-fin lobe 38 to 57% SL. Anderson and Baldwin (2000) gave a detailed description of *Anthias noeli*.

Coloration. Anderson and Baldwin (2000:377, fig. 3) presented a color illustration of one of the paratypes of *Anthias noeli*. The following description, from Anderson and Baldwin (2000:376), is based on color photographs, taken shortly after capture, of two paratypes and color notes made in the field of two other paratypes.

Head mostly rosy, a yellow streak extending across lachrymal and part of cheek and a broad yellow stripe extending from posterior margin of eye to posterior tip of opercle. Jaws rosy except small patch of yellow on upper lip near premaxillary symphysis. Iris yellow. Body mostly rosy dorsally, paler ventrally, a few yellow stripes or blotches on lateral and ventral aspects of body; black blotch present at anterior base of spinous dorsal fin. Membrane covering dorsal-fin spines yellow except rosy distally; interradial membranes mostly rosy; soft dorsal fin mostly rosy, except membranes separating three anteriormost dorsal soft rays mostly yellow or yellow distally and last four rays pale purple. Anal-fin spines and interradial membranes yellow; anterior soft anal-fin rays and interradial membranes yellow basally, yellow and rosy distally; posterior portion of soft anal fin rosy to pale purple. Pectoral fin rosy. Pelvic fin mostly pale rose, some yellow basally and on membrane between first and second rays. Caudal fin mostly rosy, some yellow on outer rays of dorsal and ventral lobes; produced distal ends of caudal-fin lobes rosy or lilac to purplish in color.

Coloration in alcohol. Anderson and Baldwin (2000:377) wrote: "Dark spot anterior to base of spinous dorsal fin usually persisting; dorsum of head frequently dusky; fins mostly pallid; remainder of fish straw-colored."

New records. Béarez and Jiménez Prado (2003) reported *Anthias noeli* from a collection made off Santa Rosa, Ecuador (02°12'S, 80°57'W), and provided, in addition to a good color photograph (fig. 5), meristic and morphometric data for three specimens (197–216 mm SL) they examined. Their data agree well with those in the original description. Béarez and Jiménez Prado (2003) gave the depth of capture of their specimens as ca. 100 meters, much shallower than the depth range (184–351 meters) reported by Anderson and Baldwin (2000), and speculated that because their specimens were collected during a cold-event La Niña that *Anthias noeli* might be a permanent resident of the mainland waters of Ecuador.

Brian Zgliczynski sent (09 May 2006, 05 March 2007) images (taken from a submersible) of *A. noeli* observed off Cocos Island. (Cocos Island is northeast of the Galápagos Islands at 05°32'N, 87°04'W.) John McCosker (*in litt.* to WDA, 06 March 2007) wrote that he observed *A. noeli* from a submersible at Cocos "along steeper slopes among and outside of the cracks in the reef face. Shallowest I saw was at 286 m, and saw several more at ~300 m." (On McCosker's dives the submersible did not venture below 300 meters.) McCosker also noted that *A. noeli* "did seem to replace, or at least not overlap, *Pronotogrammus multifasciatus*, and when I started to see *Guentherus altivelis* [Ateleopodidae] I could expect to see *noeli*."

Kristen M. Green sent (11 January 2010) several images (taken from a submersible) of anthiines observed off Cocos Island and over Las Gemelas Seamounts. (Las Gemelas is southwest of Cocos at ca. 05°05'N, 87°38'W.) One of the anthiine species depicted appears to be *Anthias noeli*.

Sexuality. Anderson and Baldwin (2000) discussed the sexuality of *Anthias noeli*, based on examination of histological sections of gonads of 15 specimens (six females, 68–139 mm SL; nine males, 86–173 mm SL, including the five largest individuals, 150–173 mm SL), and concluded that the species is protogynous.

Sexual dimorphism. Anderson and Baldwin (2000:377) reported that in specimens of *Anthias noeli*

more than about 120 mm SL, females (4 specimens, 123–139 mm SL) have shorter pelvic fins (33–36% SL vs. 43–55% SL), shorter longest dorsal soft rays (31–33% SL vs. 38–45% SL), shorter longest anal soft rays (24–28% SL vs. 28–32% SL), and shorter depressed anal-fin lengths (33–37% SL vs. 37–42% SL) than do males (5 specimens, 150–173 mm SL).

Distribution. *Anthias noeli* was described originally from specimens collected in the Galápagos Islands, in depths of 184 to 351 meters. Béarez and Jiménez Prado (2003) reported three specimens (197–216 mm SL) collected near Santa Rosa, Guayas, Ecuador, by long line in ca. 100 meters. Because those three specimens were caught at the same time (15 September 2001) and during a La Niña event, Béarez and Jiménez Prado (2003) conjectured that *A. noeli* may be a permanent inhabitant of Ecuadorian coastal waters. Based on photographs of specimens from Brian Zgliczynski and information received from John McCosker, the known range of *Anthias noeli* is further extended to include Cocos Island. Its range probably also includes Las Gemelas Seamounts (see above under **New records**).

Material examined. The only specimens examined are the 17 (62–173 mm SL) listed by Anderson and Baldwin (2000:372).

Anthias woodsi Anderson and Heemstra, 1980
Longtailed Jewelfish
Fig. 23; Tables 2–7; Map 8

Anthias woodsi Anderson and Heemstra, 1980:74, figs. 1, 2 (original description, illustrations; holotype FMNH 70709, 175 mm SL; type locality Straits of Florida, off Dry Tortugas).—Anderson, 2003:1333 (species account).

Diagnosis. A species of *Anthias* distinguishable from the other members of the genus by the following combination of characters. Soft dorsal-fin rays usually 14 (rarely 15). Lateral-line scales 42 to 48. Sum of lateral-line scales plus total gillrakers on first arch, in individual specimens, 81 to 88. Longest dorsal-fin spine (usually fourth) 12 to 17% SL; fin membrane extending into a short filament at tip of each dorsal spine, but filaments never greatly produced. Pelvic-fin length 27 to 41% SL. Caudal-fin lobes well produced. Lower jaw and midline of gular region with scales.

Description. Dorsal-fin rays X, 14 or 15 (rarely 15). Anal-fin rays III, 7 or 8 (rarely 8). Pectoral-fin rays 18 (one of 23 specimens with 16 on left side). Gillrakers 11 or 12 + 26 to 29—total 38 to 40. Lateral-line scales 42 to 48 (usually 45 or 46). Circum-caudal-peduncular scales 21 to 24 (usually 22).

Vomerine teeth in chevron-shaped patch with apex directed anteriorly. No teeth on endopterygoids or tongue. Internarial distance 9 to 21 times in snout length. Head length 36 to 41% SL. Snout length 8 to 12% SL. Orbit diameter 11 to 15% SL. Body depth at first dorsal spine 34 to 41% SL. Depressed anal-fin length 25 to 31% SL. Pelvic-fin length 27 to 41% SL (27–33% SL in specimens less than ca. 210 mm SL, 39 & 41% SL in two specimens 218 & 232 mm SL). Caudal-fin lobes extremely elongated in larger individuals. Upper caudal-fin lobe 49 to 110% SL (49–58% SL in specimens 143–175 mm SL, 63–110% SL in specimens 200–244 mm SL). Lower caudal-fin lobe 41 to 95% SL (41–56% SL in specimens 110–175 mm SL, 62–95% SL in specimens 200–244 mm SL). Anderson and Heemstra (1980) gave a detailed description of *Anthias woodsi*, and Anderson (2003) provided a brief species account.

Coloration. Steve W. Ross provided color transparencies of two specimens of *A. woodsi*; the coloration of one (UF 101424, 218 mm SL) of these specimens had faded considerably before it was photographed, as a consequence of being frozen for some time; the other (UF 101423, 232 mm SL; see Fig. 23) was photographed relatively soon after capture. The following description is based on both specimens. Head mostly rose with admixture of yellow on snout dorsolaterally; yellow bar posterior to orbit that joins yellow region on dorsolateral aspect of posterior part of head. Iris mostly reddish orange. Body mainly rose, darker immediately ventral to dorsal fin; broad yellow band beginning on opercle, narrowing ventral to posterior end of soft dorsal fin to continue over dorsolateral part of caudal peduncle—the area covered by this band pale on specimen frozen for several months, but according to Ross (*in litt.* to WDA, 23 February 1996) it "was a very bright vermillion color." Dorsal fin mostly yellow. Pectoral fin pale pink to bright rose. Anal and pelvic fins mostly pallid. Dorsalmost and ventralmost caudal-fin rays and proximal halves of filamentous portions of these rays yellow, distal halves dark rose; remainder of caudal fin mainly rose with distal band of yellow or pale yellowish green.

A color transparency of another specimen (GMBL 82-28, 166 mm SL), photographed by James F. McKinney some time after preservation, is available. Although most of the pigmentation has faded, pale red remains above lateral line and on proximal part of pectoral fin, and bright yellow is present on spinous dorsal fin, distal ends of some soft dorsal-fin rays, and at distal ends of both caudal-fin lobes.

Franklin F. Snelson, Jr., provided by e-mail attachment (10 November 2004) color photographs of two specimens of *A. woodsi* (UF 146795). One specimen has dorsal part of head, lower jaw, and midline of gular region reddish; lateral and ventrolateral aspects of head silvery; iris of eye dull yellow; body reddish near dorsal midline becoming dull orange immediately below dorsal fin, then purple along lateral line, then silvery laterally and ventrolaterally; spinous dorsal fin mostly dull yellow with tinges of red; soft dorsal fin largely pallid but distal portion mostly reddish except for yellow on anteriormost rays; anal, pectoral, and pelvic fins mostly pallid; caudal fin reddish except distal margin of fork of fin silvery and filamentous upper and lower lobes yellow. The other specimen is very similar in coloration, the most striking difference being the presence of blotches of yellow in a broad band on and ventral to lateral line extending posteriorly from opercle to a point ventral to middle of soft dorsal fin.

Ross and Quattrini (2007:993, fig. 7i) provided an *in situ* color photograph of a specimen identified by them as *Anthias woodsi*, which shows background coloration of head and body purple with yellow band (broad anteriorly) extending from above and behind eye to base of caudal fin, band narrowing posteriorly to become relatively shallow stripe on dorsolateral aspect of caudal peduncle.

Sexuality. William A. Roumillat examined histological sections of a number of specimens of *A. woodsi* and found one to be an immature female (GMBL 63-61: 110 mm SL), three to be mature females (GMBL 63-61: 143 mm SL; UF 146795: 161 & 205 mm SL), one to be a mature male (GMBL 94-12: 244 mm SL), and one (UF 146795: 200 mm SL) very clearly showing transformation from female to male. Sadovy and

Shapiro (1987:150) gave criteria for diagnosing various types of hermaphroditism in fishes. Features that they identified as strongly indicating protogyny are:

> membrane-lined central cavities in testes; transitional individuals; atretic bodies in stages 1, 2, or 3 of oocytic atresia within testes; sperm sinuses in the gonadal wall; and experimental production of transitional or sex-reversed individuals through manipulation of the social system.

Anthias woodsi meets all of those criteria except the last one (which requires study of live material) and the one involving the presence of atretic bodies within testes. The only male gonad examined was in poor condition, making it impossible to determine the presence (or absence) of old oocytes. In view of the preceding, it seems reasonable to conclude that *Anthias woodsi* is most likely protogynous.

Early life history. Baldwin (1990:929–930; fig. 11) provided a description of the early developmental stages of *A. woodsi* (tentative identification) based on four specimens, 4.0 mm NL to 14.0 mm SL, and included an illustration of a single larva (taken from Kendall, 1979, as *Anthias*—type 3), and Richards et al. (2006) presented an account of the early life history of the species.

Ecological notes. Ross and Quattrini (2007:997) observed *A. woodsi* from a submersible on deep-coral habitats off Cape Lookout, North Carolina, and wrote: "Individuals were solitary, closely associated with reef habitat, and several were observed in or darting into the *Lophelia* matrix." Those authors (2007:978) described the deep-reef habitat off North Carolina as "large mounds and ridges . . . which appear to be a sediment/ coral rubble matrix topped with almost monotypic stands (up to 3 m high) of live and dead *L. pertusa*." (*Lophelia pertusa* is a reef-building deep-water scleractinian coral.)

Distribution. *Anthias woodsi* has been collected off South Carolina, Georgia, east coast of Florida, and Dry Tortugas in depths of 174 to 475 meters. Ross and Quattrini (2007:997) reported 14 adults (ca. 300 mm total length) that they observed during dives of the Johnson-Sea-Link submersible off Cape Lookout, North Carolina, in depths of 367 to 407 meters. In late July 2009, WDA received via e-mail attachment a photograph of an anthiine caught in 90 to 183 meters of water at Norfolk Canyon off Virginia Beach, Virginia, with coloration very similar to that of the specimen of *A. woodsi* depicted in Ross and Quattrini (2007:993, fig. 7i).

Material examined. Twenty three specimens, 64 to 244 mm SL. In addition to the 12 specimens listed by Anderson and Heemstra (1980:77, 79), we have examined 11 others. **SOUTH CAROLINA:** GMBL 94-12 (1 specimen: 244 mm SL), UF 101423 (1: 232), UF 101424 (1: 218). **GEORGIA:** GMBL 89-19 (1: 152). **FLORIDA (ATLANTIC):** GMBL 82-28 (1: 166). **STRAITS OF FLORIDA OFF DRY TORTUGAS:** GMBL 63-61 (2: 110–143), UF 146795 (4: 64–205).

HOLANTHIAS GÜNTHER, 1868

Table 1

Holanthias Günther, 1868:226 (type species *Anthias fronticinctus* Günther, 1868:226, by original designation).

Diagnosis. *Holanthias* is distinguishable from all other genera of Anthiinae covered herein by the following combination of characters. Scales ctenoid, with only marginal cteni, no ctenial bases in posterior fields (Fig. 1B). Secondary squamation present at bases of larger scales. Maxilla scaly. Soft rays in dorsal fin 15 or 16. Pectoral-fin rays 19 to 21. Tubed scales in lateral line 46 to 55. No fleshy papillae on border of orbit. Vomerine tooth patch with well-developed posterior prolongation.

Description. No supramaxilla. Preopercle serrate but without antrorse spines on lower limb. Outer teeth in both jaws mostly conical, some enlarged into canines anteriorly (a few of these exserted); upper jaw with inner band of villiform to small conical teeth; anteriorly in both jaws a patch of very small teeth, one or more of these enlarged into recurved canines. Vomer, palatines, and tongue with patches of teeth; endopterygoids usually with teeth.

Single dorsal fin; not incised at junction of spinous and soft-rayed portions. Dorsal-fin rays X, 15 or 16. Anal-fin rays III, 7. Principal caudal-fin rays 15 (8 + 7); branched rays 13 (7 + 6). Gillrakers well developed, total on first arch 38 to 43. Vertebrae 26 (10 + 16). First caudal vertebra without parapophyses. Formula for configuration of supraneural bones, etc. 0/0/2/1 + 1/1/ (Fig. 2B). Pleural ribs on vertebrae 3 through 10. No trisegmental pterygiophores associated with dorsal and anal fins.

Lateral line complete, not interrupted, extending to at least base of caudal fin, running approximately parallel to dorsal body contour a few scale rows ventral to dorsal-fin base, then curving ventral to posterior end of dorsal-fin base to continue near midlateral axis of caudal peduncle. Most of head, including maxilla, dorsum of snout, and interorbital region with scales. Dorsal and anal fins mostly without scales, but with some scales basally; pectoral, pelvic, and caudal fins with scales basally. Head with three brightly colored stripes, one running just ventral to eye (or near ventral border of eye), one just dorsal to eye (or near dorsal border of eye), and one in the occipital region. (Much of the preceding is based only on *Holanthias fronticinctus* because we have not examined any specimens of *H. caudalis*.)

Remarks. Although our understanding of the generic classification of anthiine fishes is still very incomplete, it seems clear that *H. fronticinctus* and *H. caudalis* are generically distinct from the other anthiines reported herein. Various workers have considered *Odontanthias* Bleeker, 1873, *Ocyanthias* Jordan and Evermann, 1896, and *Scalantarus* Smith, 1965, as junior synonyms of *Holanthias*. Randall and Heemstra (2006) treated *Odontanthias* as valid and *Scalantarus* as a junior synonym of it. Herein we regard *Ocyanthias* as a junior synonym of *Pronotogrammus*.

We have not examined any specimens of *H. caudalis*, but judging from the original description, it is very similar to *H. fronticinctus*.

Key to the Species of *Holanthias*

1a. Lateral-line scales 46 or 47; two rays of lower caudal-fin lobe greatly elongated .... *Holanthias caudalis*
(central South Atlantic: southeast of Ascension Island)

1b. Lateral-line scales 50–55; posterior margin of caudal fin convex with middle of fin angulated in some specimens; middle rays of caudal fin elongated in some larger individuals .... *Holanthias fronticinctus*
(eastern South Atlantic: off Island of Saint Helena)

Holanthias caudalis Trunov, 1976

Whiptail Seaperch

Tables 2–6; Map 9

Holanthias caudalis Trunov, 1976:230, figs. 1–3 (original description; illustrations; types AtlantNIRO Museum No. 33, holotype 214 mm SL, paratype 195 mm SL; type locality southeast of Ascension Island).—Anderson, in press (species account).

Diagnosis. *Holanthias caudalis* can be distinguished from the only other species in the genus *Holanthias, H. fronticinctus,* by the following. Lateral-line scales 46 or 47. Two rays of lower caudal-fin lobe greatly elongated.

Description. Dorsal-fin rays X, 15. Anal-fin rays III, 7. Pectoral-fin rays 21. Gillrakers 12 + 29 or 30—total 41 or 42. Trunov (1976:230) stated: "plates of small teeth on the vomer, the palatine bones and the tongue," and his fig. 2 shows the vomerine tooth patch with a well-developed posterior prolongation, the palatines and tongue with well-developed patches of teeth, and a small patch of teeth on each endopterygoid. Head length 28.0 to 28.7% SL. Snout length 7.9 to 8.1% SL. Eye diameter 6.5% SL (in holotype and paratype). Body depth 37.0 to 38.8% SL. Length of third dorsal-fin spine 13.8 to 14.0% SL. Length of longest soft ray in dorsal fin (eleventh) 14.6 to 20.3% SL. Second soft ray in pelvic fin longest, third soft ray in anal fin longest, both produced in fig. 1 of Trunov (1976), as is 11th dorsal soft ray. Two rays of lower caudal-fin lobe greatly elongated. The preceding data are from Trunov (1976). Anderson (in press) provided a brief species account.

Coloration. Trunov (1976:232) presented the following description of the coloration of *H. caudalis.*

The overall body color is pinkish-yellow. There are 3 clear rose-colored stripes on the head (infraorbital, supraorbital and occipital). The rest of the head and also the pelvic and partly the anal fins are rose-colored or pinkish. There is a narrow bright rose-colored border along the base and at the ends of the rays of the dorsal fin (its inner part is yellow). The caudal and anterior margins of the anal fin are similarly edged. Moreover, on the anal fin there is a transversely linear, comparatively wide yellow band. The remaining part of the body is pinkish with small spots of a more intensive rose scattered over its surface.

Remarks. Allué et al. (2000:110) reported two specimens of *Holanthias caudalis* collected from the eastern South Atlantic north of the Island of Saint Helena. The first author examined both specimens and identified them as *Anthias helenensis*. Although we have not examined any specimens of *H. caudalis*, this species appears to be distinct from *H. fronticinctus* in having fewer lateral-line scales, a differently shaped caudal fin, and an elongated 11th dorsal soft ray.

Distribution. *Holanthias caudalis* was described from two specimens collected southeast of Ascension Island (central South Atlantic Ocean) in 120 to 170 meters of water.

Material examined. None.

Holanthias fronticinctus (Günther, 1868)
Deepwater Greenfish
Fig. 24; Tables 2–7; Map 9

Anthias fronticinctus Günther, 1868:226, pl. 18 (original description; illustration; lectotype, herein designated, BMNH 1867.10.8.40, 173 mm SL; type locality off the Island of Saint Helena, South Atlantic Ocean).

Holanthias fronticinctus: Anderson, in press (species account).

Diagnosis. *Holanthias fronticinctus* can be distinguished from the only other species of *Holanthias, H. caudalis,* by the following. Lateral-line scales 50 to 55. Posterior margin of caudal fin convex with middle of fin angulated in some specimens; middle rays of caudal fin elongated in some larger individuals.

Description. Dorsal-fin rays X, 15 or 16 (usually 15). Anal-fin rays III, 7. Pectoral-fin rays 19 to 21 (usually 20). Gillrakers 10 to 14 + 26 to 30—total 38 to 43. Circumcaudal-peduncular scales 24 to 26. Procurrent caudal-fin rays 10 or 11 dorsally, 10 to 12 ventrally. Epineurals associated with first 11 vertebrae.

Nares closely set on each side of snout; internarial distance 7 to 13 times in snout length. Head length 28 to 30% SL. Snout length 6 to 8% SL. Orbit diameter 7 to 8% SL. Body depth at first dorsal spine 34 to 39% SL. Longest dorsal spine (third) 11 to 14% SL. Depressed anal-fin length 25 to >39% SL. Pelvic-fin length 28 to 55% SL. Length of midcaudal fin rays 27 to 44% SL. (See Edwards and Glass, 1987:636, for comparisons of males and females in fin and fin-ray lengths.) Anderson (in press) gave a brief species account.

Coloration. Edwards and Glass (1987:636) wrote that *H. fronticinctus* has:

a brilliant yellow-orange colour in life with a purple band along the base of, and purple edging to the dorsal fin. On the head a purple band runs obliquely from the snout through the lower edge of the orbit to the preoperculum and a second, parallel, intermittent purple band passes from the forehead through the upper edge of the orbit to the preoperculum. A short purple band is also present across the nape, and the pectoral axil is purple.

In preserved specimens locations of the purple bands on head and nape are usually clearly indicated by white (or light) bands. Edwards (1990: pl. 11.2) presented a color photograph of a female (ca. 200 mm SL) of *H. fronticinctus*.

Designation of lectotype. There are two syntypes of *Anthias fronticinctus*: BMNH 1867.10.8.40 (173 mm SL) and BMNH 1867.12.26.16 (174 mm SL). To firmly associate the name with a specimen, we designate BMNH 1867.10.8.40 as the lectotype of *Anthias fronticinctus*. The other specimen, BMNH 1867.12.26.16, becomes the paralectotype.

Sexuality and sexual dimorphism. Edwards and Glass (1987:636) wrote that *H. fronticinctus* "appears to undergo sex reversal, females changing to males at *ca.* 204–212 mm SL." They examined the gonads of several specimens histologically and found three specimens (212–218 mm SL) with mature testes and one of 190 mm SL with eggs. Edwards and Glass (1987:636) also reported females of 172 to 190 mm SL with length of caudal fin as 26.3 to 27.6% SL, length of second pelvic soft ray as 22.2 to 37.7% SL, and length of third anal soft ray as 15.3 to 20.5% SL. In the two largest males (213 & 218 mm SL) they examined, data for those measurements are: caudal fin 36.0 to 42.7% SL, second pelvic soft ray 47.8 to 53.7% SL, and third anal soft ray 25.9 to 29.4% SL.

Distribution. *Holanthias fronticinctus* is known from waters off the Island of Saint Helena in the eastern South Atlantic Ocean in depths of 73 to 110 meters. Edwards (1993, based on Hoogesteger, 1988) noted that up to 20% of the diet of *Thunnus albacares* (Yellowfin Tuna) caught at Bonaparte Seamount (about 130 kilometers west of Saint Helena) is made up of *Holanthias fronticinctus*. Unfortunately, no specimens are available to verify this locality record.

Material examined. Ten specimens, 170 to 216 mm SL. **OFF THE ISLAND OF SAINT HELENA:** BMNH 1867.10.8.40 (lectotype: 173 mm SL), BMNH 1867.12. 26.16 (paralectotype: 174), BMNH 1969.11.19.1 (1: 180), BMNH 1984.7.16.47–49 (3: 189–208), BMNH 1984.7.16.50 (1: 216), BMNH 1985.12.6:13 (1: 170), SAM 24615 (1: 193), USNM 267911 (1: 203).

MEGANTHIAS RANDALL AND HEEMSTRA, 2006

Table 1

Meganthias Randall and Heemstra, 2006:27 (type species *Sacura natalensis* Fowler, 1925:226, by original designation).

Diagnosis. Species of *Meganthias* included in this work are distinguishable from species of all other genera of Anthiinae treated herein by the following combination of characters. Scales ctenoid, with only marginal cteni, no ctenial bases in posterior fields (Fig. 1B). Secondary squamation present at bases of larger scales. Maxilla scaly. Soft rays in dorsal fin 17 or 18. Pectoral-fin rays 16 or 17. Lateral-line scales 46 to ca. 50. No fleshy papillae on border of orbit. Vomerine tooth patch without posterior prolongation.

Description. This description is limited to the two species of *Meganthias* considered herein. Mouth terminal, oblique, lower jaw exceeding upper jaw slightly with mouth closed. Nares very closely set on each side of snout. No supramaxilla. Preopercle serrate but without spine at angle or antrorse spines on lower limb. Premaxilla and dentary each with outer series of small conical teeth (some enlarged anteriorly) and inner band of villiform teeth. Vomer, palatines, and tongue with very small teeth; endopterygoids without teeth. Vomerine tooth patch without posterior prolongation, but apex of patch with one to several enlarged conical or canine teeth.

Single dorsal fin; not incised at junction of spinous and soft-rayed portions; no elongated dorsal spines; two to several soft rays in dorsal fin produced. Dorsal-fin rays X, 17 or 18. Anal-fin rays III, 8. Caudal fin lunate to forked; principal rays 15 (8 + 7); branched rays 13 (7 + 6); procurrent rays 6 or 7 dorsally, 7 ventrally. Gillrakers well developed, total on first arch 35 to 39. Vertebrae 26 (10 + 16). First caudal vertebra without parapophyses. Formula for configuration of supraneural bones, etc. 0/0/2/1+1/1/ (Fig. 2B). Pleural ribs on vertebrae 3 through 10. Epineurals associated with first 11 to 13 vertebrae. No trisegmental pterygiophores associated with dorsal and anal fins.

Remarks. *Meganthias* includes four described species—*M. carpenteri* Anderson, 2006, from the eastern Atlantic; *M. natalensis* (Fowler, 1925) from the western Indian Ocean; *M. filiferus* Randall and Heemstra, 2008, from the Andaman Sea (eastern Indian Ocean) and Arabian Sea off the southwest coast of India (Akhilesh et al., 2009); and *M. kingyo* (Kon, Yoshino, and Sakurai, 2000) from the Ryukyu Islands in the western Pacific. A specimen collected at Vema Seamount in the eastern South Atlantic appears to represent a fifth species, and we treat it as such herein as *Meganthias* sp.

Key to Atlantic *Meganthias*

1a. Lips mostly covered with very small scales, upper lip without rugosities, lower lip with rugosities along dorsal aspect; maxilla not broadly rounded posterodorsally; total gillrakers on first arch 35–37; pectoral-fin length 34% SL; caudal-peduncle length 18–21% SL; tenth-dorsal spine length 14% SL; bony interorbital width 11% SL . . . . *Meganthias carpenteri* (eastern Atlantic: off Nigeria)

1b. Lips rugose and with very small scales; maxilla broadly rounded posterodorsally; total gillrakers on first arch 39; pectoral-fin length 31% SL; caudal-peduncle length 23% SL; tenth-dorsal spine length 12% SL; bony interorbital width 13% SL . . . . *Meganthias* sp. (eastern South Atlantic: Vema Seamount)

Meganthias carpenteri Anderson, 2006

Yellowtop Jewelfish

Fig. 25; Tables 2–7; Map 9

Meganthias carpenteri Anderson, 2006:405, figs. 1–3 (original description; illustrations; holotype USNM 386079, 301 mm SL; type locality off Nigeria, eastern Atlantic ocean).—Anderson, in press (species account).

Diagnosis. *Meganthias carpenteri* can be distinguished from the only other Atlantic species in the genus *Meganthias, Meganthias* sp., by the following combination of characters. Lips mostly covered with very small scales, upper lip without rugosities, lower lip with rugosities along dorsal aspect. Maxilla not broadly rounded posterodorsally, posterior border of maxilla almost a straight line. Total gillrakers on first arch 35 to 37. Pectoral-fin length 33.5 to 33.7% SL. Caudal-peduncle length 18.2 to 20.9% SL. Tenth dorsal-spine length 13.6 to 14.3% SL. Bony interorbital width 10.6 to 11.2% SL.

Description. Dorsal-fin rays X, 17 or 18. Anal-fin rays III, 8. Pectoral-fin rays 16 or 17. Gillrakers 11 or 12 + 24 or 25—total 35 to 37. Lateral-line scales 46 to ca. 50. Circum-caudal-peduncular scales ca. 25 or 26 (difficult to count).

Most of head, including lips, most of dorsal and lateral aspects of snout, lachrymal region, interorbital region, maxilla, and lower jaw covered with scales. Branchiostegal rays, branchiostegal membranes, and gular region naked (except holotype with patch of scales in middle of gular region and a few scales on two branchiostegal rays). Soft dorsal, anal, pectoral, pelvic, and caudal fins densely covered with scales basally and with scales extending distally well out on interradial membranes. Axillary scales poorly developed. Scales in ventral midline between pelvic fins (interpelvic process) well developed.

Nares very closely set on each side of snout; internarial distance 14 to 17 times in snout length. Head length 34.6 to 36.0% SL. Snout length 8.0 to 8.4% SL. Orbit diameter 7 to 9% SL. Body depth at first dorsal spine 45 to 48% SL. Longest dorsal spine (fourth or fifth) 14 to 16% SL; fin membrane extending into a short filament posterior to distal tip of each dorsal spine, but filaments never greatly produced. Two to four anterior soft rays in dorsal fin produced, longest dorsal soft ray ca. 33 to ca. 62% SL. Depressed anal-fin length 32 to 34% SL. Pelvic-fin length 30 to 35% SL. Caudal fin lobes damaged on two known specimens, but apparently were long, slender, and pointed; lower lobe of one specimen only slightly damaged, >44.5% SL. Anderson (2006) provided a detailed description of *Meganthias carpenteri*, and Anderson (in press) gave a short species account.

Coloration. Anderson (2006:407, fig. 2) presented a color illustration of the holotype and paratype of *Meganthias carpenteri* (the only known specimens of the species). The following description, from Anderson (2006:408), is based on color photographs of the specimens taken after thawing, but before preservation.

> Head mostly rosy, but with bright yellow on much of snout, lachrymal region, and anterior part of preopercle, and bordering maxilla, dentary, and ventral margin of orbit. Bright yellow oblong area on dorsum extending from level of posterior part of orbit to at least middle of spinous dorsal fin. Iris of eye mostly yellow, with some rose peripherally. Body mostly rosy dorsally, paler ventrally; numerous yellow blotches present immediately ventral to soft dorsal fin and on caudal peduncle of holotype (yellow blotches present dorsal to base of pectoral fin and on caudal peduncle of paratype). Fins almost completely yellow, except soft dorsal fin with much rose overlain with numerous yellow streaks and spots and base of anal fin rosy. In alcohol, four months after preservation, coloration mostly faded, but yellow oblong area on dorsum still evident, though faint.

Sexuality and sexual dimorphism. Although histological examination of the gonads of the two known specimens of *Meganthias carpenteri* showed the holotype (301 mm SL) to be a male and the paratype (244 mm SL) to be a female, the specimens were frozen before preservation, resulting in the disruption of the gonadal tissue to the point where it is impossible to determine the type of sexuality present in this species.

Sexual dimorphism is evident in the development of some of the anterior soft rays in the dorsal fin. The male (301 mm SL) has the second through the fifth dorsal soft rays produced—third 57% SL, fourth ca. 62% SL. This contrasts with the female (244 mm SL), which has anterior dorsal soft rays much less elongated—second >27% SL, third >32.7% SL (tips of both rays broken).

Distribution. *Meganthias carpenteri* is only known from the eastern Atlantic Ocean off the coast of Nigeria. No data are available on depth for this species.

Material examined. Known only from the two type specimens, 244 mm SL (USNM 386080, female, paratype) and 301 mm SL (USNM 386079, male, holotype), collected off Nigeria in the eastern Atlantic Ocean.

Meganthias sp.

Vema Jewelfish

Tables 2–6; Map 9

Meganthias sp.: Anderson, 2006:411, fig. 3C (description of SAM 25287, 303 mm SL; illustration of maxilla; Vema Seamount, eastern South Atlantic Ocean).

Diagnosis. *Meganthias* sp. can be distinguished from the only other Atlantic species in the genus *Meganthias, Meganthias carpenteri*, by the following combination of characters. Lips rugose, with very small scales. Posterodorsal border of maxilla broadly rounded. Total gillrakers on first arch 39. Pectoral-fin length 30.7% SL. Caudal-peduncle length 23.1% SL. Tenth-dorsal spine length 11.6% SL; bony interorbital width 12.9% SL.

Description. Dorsal-fin rays X, 18. Anal-fin rays III, 8. Pectoral-fin rays 17. Gillrakers 11 + 28—total 39. Lateral-line scales ca. 47. Head length 35.3% SL. Snout length 9.6% SL. Orbit diameter 8.1% SL. Body depth at first dorsal spine 46.9% SL. No elongated spines in dorsal fin, third spine 13.4% SL. Second and third soft rays in dorsal fin well produced, third soft ray 49.8% SL. Depressed anal-fin length 33.0% SL. Pelvic-fin length 29.7% SL. Caudal fin forked (both lobes cut off). Anderson (2006) provided a short description of *Meganthias* sp.

Remarks. Anderson (2006:411–412) discussed the apparent affinities of *Meganthias* sp. from Vema Seamount and concluded that it seems more similar to *M. natalensis* from the southwestern Indian Ocean than to *M. carpenteri* from the eastern Atlantic but was "inclined to think that the Vema individual represents a species distinct from *M. natalensis.*"

Distribution. *Meganthias* sp. is known only from Vema Seamount (31°38'S, 08°20'E) in the eastern South Atlantic Ocean. No depth data are available, but the summit of Vema Seamount is within 26 meters of the surface (Berrisford, 1969).

Material examined. Known from a single specimen (SAM 25287, 303 mm SL) collected over Vema Seamount (Anderson, 2006).

ODONTANTHIAS BLEEKER, 1873

Table 1

Odontanthias Bleeker, 1873:236 (type species *Serranus borbonius* Valenciennes, 1828:263, by original designation).

Scalantarus Smith, 1965:535 (type species *Anthias chrysostictus* Günther, 1872:655, by original designation (also monotypic).

Diagnosis. The single species of *Odontanthias* included herein is distinguishable from all other species treated in this study by the following combination of characters. Scales ctenoid, with only marginal cteni, no ctenial bases in posterior fields (Fig. 1B). No secondary squamation on body, but a few accessory scales present on head. Maxilla scaly. Soft rays in dorsal fin 15. Soft rays in anal fin 7. Pectoral-fin rays 18. Total number of gillrakers on first arch 42 or 43. Lateral-line scales 33 to 38. Circum-caudal-peduncular scales 16 to 18. No fleshy papillae on border of orbit. Vomerine tooth patch not prolonged posteriorly. Caudal fin lunate with very long filamentous lobes.

Description. The description is limited to the single species of *Odontanthias* considered herein. For that description see below under *Odontanthias hensleyi*.

Remarks. Randall and Heemstra (2006) reviewed the genus *Odontanthias* Bleeker, 1873, considering it to include 13 Indo-Pacific species, two of which were new to science, and described a new anthiine genus, *Meganthias*. Based on morphology the genera most closely related to *Odontanthias* would appear to be *Holanthias* (with which it has been frequently synonymized) and *Meganthias* (described by Randall and Heemstra, 2006). Randall and Heemstra (2006:4) distinguished *Odontanthias* from *Holanthias* on the basis of the shape of the caudal fin ("deeply emarginate with rounded lobes to lunate with slender, sometimes filamentous lobes" in *Odontanthias* vs. "near-truncate to rounded or rhomboid . . . with a long slender lobe in the ventral part of the fin of one of the species" in *Holanthias*) and the absence of accessory scales on the body scales of species of *Odontanthias* (although present "on the head and nape of a few species") vs. "numerous accessory scales on the body scales of *Holanthias*."

In the diagnosis presented in the original description of *Meganthias*, Randall and Heemstra (2006) wrote that their new genus had the characters of *Odontanthias* except for a number of morphological differences, the most important of which would seem to be anal soft rays 8 or 9 and the presence of accessory scales ("dense on head and nape," p. 27) in *Meganthias* vs. anal soft rays 7 or 8 (usually 7) and the absence of accessory scales on the body ("but may be present on head and nape of some species," p. 4) in *Odontanthias*.

The additions of *Odontanthias hensleyi* and a new species (*O. randalli),* described from Indonesian waters by White (2011), brings the number of species in the genus to 15.

Odontanthias hensleyi Anderson and García-Moliner, 2012

Euripos Jewelfish

Fig. 26; Tables 2–7; Map 6

Odontanthias hensleyi Anderson and García-Moliner, 2012:26, Fig. 1 (original
   description; illustration; holotype USNM 400888, 155 mm SL; type locality
   western North Atlantic, Mona Passage, off west coast of Puerto Rico—18°07'N,
   67°40'W).

Diagnosis. As for the genus.

Description. Dorsal-fin rays X, 15. Anal-fin rays III, 7. Pectoral-fin rays 18. Caudal-fin rays: principal 15 (8 + 7); branched 13 (7 + 6); procurrent 8 dorsally, 8 ventrally. Gillrakers on first arch 13 or 14 + 28 to 30—total 42 or 43. Lateral-line scales 33 to 38. Circum-caudal-peduncular scales 16 to 18. Vertebrae 26 (10 + 16). First caudal vertebra without parapophyses. Formula for configuration of supraneural bones, etc. 0/0/2/1 + 1/1/ (Fig. 2B). Pleural ribs on vertebrae 3 through 10. Epineurals associated with first 11 vertebrae. No trisegmental pterygiophores associated with dorsal or anal fins.

Head length 35 to 36% SL. Snout length 7 to 8% SL. Orbit diameter 11 to 12% SL. Body depth at first dorsal spine 39 to 42% SL. Dorsal and anal spines without long filaments. Dorsal spines not produced; longest dorsal spine (fourth, fifth, ninth, or tenth) 11 to 14% SL. Depressed anal-fin length 34 to 38% SL. First three pelvic soft rays produced, second longest—reaching past base of anal fin to well past base of caudal fin; pelvic-fin length 61 to 80% SL. Upper lobe of caudal fin 2.1 to 2.4 times length of head (>73 to 86% SL); lower lobe of caudal fin 1.9 to 2.3 times length of head (>66 to 85% SL).

Mouth terminal, oblique; premaxillae protrusile; lower jaw usually exceeding upper jaw slightly with mouth closed. Supramaxilla apparently present (due to dense squamation on upper jaw, presence difficult to determine with certainty). Nares closely set on each side of snout; internarial distance 7 to 13 times in snout length; anterior naris in short tube, posterior border of tube not reaching posterior naris when reflected. Preopercle serrate but without spine at angle or antrorse spines on lower limb.

Premaxilla with series of conical teeth laterally and band of villiform teeth medially; at anterior end of jaw one or two canine(s) adjacent to patch of very small teeth; symphysis edentate. Dentary with 1 to 3 recurved canine(s) about one-fourth to one-third way back from anterior end of jaw; anterior to recurved canine(s) patch or band of villiform to very small conical teeth; posterior to recurved canine(s) series or band of small conical teeth extending along jaw; exserted canine at anterior end of jaw; symphysis edentate. Vomer, palatines, endopterygoids, and tongue with small teeth; vomerine tooth patch subquadrangular to diamond shaped, without posterior prolongation; palatine and endopterygoid teeth in longitudinal patches (endopterygoid teeth not seen on two of four known specimens).

Lateral line complete, anteriorly ascending above pectoral-fin base to run parallel to dorsal body contour a few scale rows ventral to base of dorsal fin, then descending precipitously ventral to posterior end of soft dorsal fin to run posteriorly near middle of caudal peduncle to terminate at distal end of hypural bones. Most of head, including dorsum of snout, interorbital region, maxilla, and dentary covered with scales. Lips, lateral aspect of snout, lachrymal, gular region, branchiostegals, and branchiostegal membranes without scales. Spinous dorsal fin without scales; soft dorsal and anal fins with scales basally and with columns of scales on some of the interradial membranes; pectoral and pelvic fins scaly basally and for some distance out onto fins; most of caudal fin heavily covered with scales. Pelvic axillary scales poorly developed or absent; scales in ventral midline between pelvic-fin bases (interpelvic process) well developed.

The above description is a modification of the description in Anderson and García-Moliner (2012).

Coloration. Anderson and García-Moliner (2012) provided a detailed description of the coloration of *O. hensleyi* (see Fig. 26 for color photograph of the holotype). Striking features of this species are side of head with two wavy bright yellow stripes and pelvic, anal, and caudal fins bright yellow.

Comparisons.—Based on data and illustrations provided by Randall and Heemstra (2006), *O. hensleyi* differs from most other species of *Odontanthias* in having 15 soft rays in the dorsal fin, whereas *caudicinctus*, *dorsomaculatus*, *elizabethae*, *grahami*, *rhodopeplus*, and *unimaculatus* have 14 or fewer, and *borbonius*, *flagris*, *fuscipinnis*, *katayamai*, and *wassi* have 16 or more, and it differs from the other two species of the genus considered by Randall and Heemstra (2006) in having 18 pectoral-fin rays, whereas they, *chrysostictus* and *tapui*, have 17 or fewer. *Odontanthias randalli* White, 2011, from Indonesian waters, has 16 or 17 soft rays in the dorsal fin and 15 or 16 pectoral-fin rays. Also, *O. hensleyi* differs from most other *Odontanthias* in having the vomerine teeth in a subquadrangular to diamond-shaped patch without a posterior prolongation, in contrast to a variety of shapes in those other species (see Randall and Heemstra, 2006:8, fig. 1; White, 2011, fig. 2). In addition *O. hensleyi* can be distinguished from other *Odontanthias* by the following combination of characters: pelvic fin reaching past base of anal fin to well past base of caudal fin; upper and lower lobes of caudal fin produced into long filaments; head with two bright yellow stripes on side, and pelvic, anal, and caudal fins bright yellow (compare Fig. 26, herein, with figs. 4 and 5 and plates I–VI in Randall and Heemstra, 2006, and with fig. 1 in White, 2011).

Sexuality. Histological examination of gonadal tissue shows the holotype of *O. hensleyi* to be a male.

Distribution. All specimens of *O. hensleyi* were collected in Mona Passage, the strait between Puerto Rico and the Dominican Republic leading from the open Atlantic to the Caribbean Sea in depths of 338 to 344 meters.

Remarks. The only known locality for another relatively recently described fish species is Mona Passage. *Symphysanodon mona* was described from a single specimen collected by the R/V OREGON in October 1959 @18°13'N, 67°20'W, in 384 meters (Anderson and Springer, 2005). The apparent restricted distributions of these two species are probably collecting artifacts.

Material examined. Four specimens, 155 to 162 mm SL. **Mona Passage off west coast of Puerto Rico:** UPRM 3793 (1 specimen: 162 mm SL), UPRM 3794 (1: 159), UPRM 3809 (1: 157), USNM 400888 (holotype: 155).

PRONOTOGRAMMUS GILL, 1863

Table 1

Pronotogrammus Gill, 1863:80 (type species *Pronotogrammus multifasciatus* Gill, 1863:81, by monotypy).

Ocyanthias Jordan and Evermann, 1896:1131, 1227 (type species *Aylopon martinicensis* Guichenot, 1868:85, by original designation).

Pacificogramma Kharin, 1983b:116 (type species *Pacificogramma stepanenkoi* Kharin, 1983b:117, by original designation).—Anderson and Rosenblatt, 1989:124 (placed *Pacificogramma* Kharin, 1983b, in the synonymy of *Pronotogrammus* Gill, 1863, as a consequence of subsuming the type species of *Pacificogramma* into the synonymy of *Pronotogrammus multifasciatus*).

Diagnosis. *Pronotogrammus* is distinguishable from all other genera of Anthiinae covered herein by the following combination of characters. Scales ctenoid, with only marginal cteni, no ctenial bases in posterior fields (Fig. 1B). No secondary squamation (except one specimen of *P. multifasciatus* of 152 mm SL with squamulae in nape region). Maxilla with scales. Lateral line complete (occasionally interrupted in *P. multifasciatus*), tubed scales 35 to 57. Circum-caudal-peduncular scales 18 to 28. No fleshy papillae on border of orbit. Anterior and posterior nares relatively close together; posterior border of anterior naris produced into slender filament (usually falling well short of orbit when reflected). Vomerine tooth patch usually with posterior prolongation. Baldwin (1990:918, 920—fig. 4B, 950) reported a single character, type B larval scales, that is apparently a synapomorphy for the species of this genus.

Description. Premaxillae protrusile. No supramaxilla. Preopercle serrate but without antrorse spines on lower limb. Outer teeth in jaws mostly conical; inner teeth mostly villiform or cardiform; some teeth enlarged as canines. Vomer and palatine with teeth; vomerine tooth patch usually with posterior prolongation. Endopterygoid usually toothless. Tongue usually with teeth.

Single dorsal fin (not incised at junction of spinous and soft-rayed portions), dorsal-fin rays X, 13 to 16 (very rarely XI, usually 15). Anal-fin rays III, 7 or 8 (usually 7). Pectoral fin symmetrical, with 16 to 21 rays. Caudal fin lunate to moderately forked; principal rays 15 (8 + 7); branched rays 13 (7 + 6); procurrent rays 8 to 12 dorsally and ventrally. Gillrakers well developed, total on first arch 34 to 41. Vertebrae 26 (10 + 16). First caudal vertebra without parapophyses. Formula for configuration of supraneural bones, etc. 0/0/2/1+1/1/ (Fig. 2B), except *P. martinicensis* rarely with slightly different placement of one of the supraneural bones. Pleural ribs on vertebrae 3 through 10. Epineurals associated with first 10 to 12 vertebrae. No trisegmental pterygiophores associated with dorsal and anal fins.

Lateral line usually complete (infrequently disjunct in *P. multifasciatus*), extending to at least base of caudal fin, running parallel to dorsal body contour a few scale rows below dorsal fin, then descending rather steeply beneath posterior end of dorsal-fin base to continue near midlateral axis of caudal peduncle. Most of head, including maxilla, dorsum of snout, interorbital region, and ventral surface of lower jaw scaly.

Branchiostegals and branchiostegal membranes without scales. Dorsal and anal fins without scales or mostly without scales. Pectoral, pelvic, and caudal fins scaly basally and for varying distances out onto fins.

Remarks. Baldwin (1990:950) wrote that *"Pronotogrammus martinicensis . . . and P. multifasciatus* appear to be sister species" that share "three derived features." Those features are: type B larval scales (described by Baldwin, 1990), posterior prolongation of the vomerine tooth patch, and an ovate lingual tooth patch. Two of those characters are not reliable indicators of relationships. About 40% of the specimens of *P. multifasciatus* examined lack a backward prolongation of the vomerine tooth patch, and about 30% lack lingual teeth; in those with lingual teeth the patches are variously shaped.

Anderson and Heemstra (1980) pointed out that both *Pronotogrammus martinicensis* (as *Holanthias martinicensis*) and *Choranthias tenuis* (as *Anthias tenuis*) have the posterior border of the anterior naris produced into a slender filament. Although acknowledging that the produced narial filament could have been derived independently in those two species, they wrote that its presence might indicate close relationship. Baldwin (1990), in her study of the larvae of Atlantic and eastern Pacific Anthiinae, could find no other characters suggesting that *C. tenuis* and *P. martinicensis* are closely related (see **Remarks** under *Choranthias*). We agree with her interpretation that the produced narial filaments probably evolved independently in two distinct lineages of anthiines. Consequently, we consider the presence of a filament on the posterior border of the anterior naris to be a synapomorphy uniting the species of *Pronotogrammus*.

Key to the Species of *Pronotogrammus*

1a. Lateral-line scales 35–43 (usually 38–40); pectoral-fin rays 16–18 (usually 17); circum-caudal-peduncular scales 18–23 (usually 20 or 21); longest dorsal-fin spine (usually the third) 16–25% SL . . . . *Pronotogrammus martinicensis* (western Atlantic)

1b. Lateral-line scales 46–57 (usually 48–54); pectoral-fin rays 19–21 (usually 19 or 20); circum-caudal-peduncular scales 23–28; longest dorsal-fin spine (most frequently the fourth) 11 to >17% SL, 11–13% SL in specimens more than ca. 170 mm SL . . . . *Pronotogrammus multifasciatus* (eastern Pacific)

Pronotogrammus martinicensis (Guichenot, 1868)

Roughtongue Bass

Fig. 27; Tables 2–7; Map 10

Aylopon martinicensis Guichenot, 1868:85 (original description; lectotype MNHN 4319, 97 mm SL; type locality off Martinique, French West Indies).

Anthias duplicidentatus Miranda Ribeiro, 1903:26 (original description; holotype apparently lost; type locality "ESE. da ilha Rasa," Brazil).

Anthias louisi Bean, 1912:124 (original description; holotype AMNH 7308, 80 mm SL; type locality Argus Bank, Bermuda).

Ocyanthias martinicensis: combination used by various authors.

Holanthias martinicensis: combination used by various authors, including Anderson and Heemstra, 1980:82 (species account).

Pronotogrammus martinicensis: Anderson, 2003:1364 (species account).

Diagnosis. *Pronotogrammus martinicensis* can be distinguished from the only other species in the genus *Pronotogrammus, P. multifasciatus*, by the following combination of characters. Lateral-line scales 35 to 43 (usually 38–40). Pectoral-fin rays 16 to 18 (usually 17). Circum-caudal-peduncular scales 18 to 23 (usually 20 or 21). Longest dorsal-fin spine (usually the third) 16 to 25% SL.

Description. Dorsal-fin rays X, 13 to 16 (usually 15). Anal-fin rays III, 7 (very rarely 8). Gillrakers 9 to 13 + 24 to 29—total 34 to 41 (usually 36–39). Procurrent caudal-fin rays 8 to 10 dorsally and ventrally. Epineurals associated with first 11 or 12 vertebrae.

Vomerine tooth patch with well-developed posterior prolongation. Endopterygoids with teeth in 20% of specimens examined. Tongue with large oval patch of teeth. Internarial distance 5 to 8 times in snout length. Head length 31 to 43% SL (31–37% SL in specimens more than ca. 50 mm SL). Snout length 5 to 9% SL. Orbit diameter 9 to 16% SL (9–13% SL in specimens more than ca. 75 mm SL). Body depth at first dorsal spine 33 to 42% SL. Depressed anal-fin length 30 to 39% SL. Pelvic-fin length 25 to 35% SL. Caudal fin crescentic; outer principal rays of both lobes somewhat produced. Upper caudal-fin lobe 28 to >51% SL. Lower caudal-fin lobe 28 to >58% SL. Anderson and Heemstra (1980) presented a species account under the name *Holanthias martinicensis*, and Anderson (2003) presented a short species account.

Coloration. Dennis and Bright (1988b:6–7) commented briefly on the in situ coloration of *Pronotogrammus martinicensis* (as *Holanthias martinicensis*), based on examination of Kodachrome transparencies. They noted "a blue hue to the pelvic and anal fins and a distinctive dark bar (yellow if the light source was close enough) across the mid-body just behind the pelvic fins." Bullock and Smith (1991:209, pl. II, fig. C) presented a color photograph of a 73-mm specimen (FSBC 18002) of this species; their figure shows a mainly dull orange individual with yellow stripe from near anterior tip of snout (running below eye) to posterior margin of opercle. Coleman (1981:894) noted that "many preserved females have a dark bar beneath the sixth through ninth dorsal-fin spines, extending ventrally to horizontal from the posteriormost point of the preopercle." Matsuura (1983:312) published a color photograph of this species (as *Ocyanthias martinicensis*) that shows a prominent yellow bar at midbody (from spinous dorsal fin to near ventral midline) and described the coloration as: "Body pale yellowish olive, anterior half of body yellow; fins pale yellow."

A color transparency, received from Donald D. Flescher, of a specimen of *P. martinicensis* of 112 mm SL collected off South Carolina (GMBL 81–150) shows the following: Head red orange dorsally; cheek rosy; oblique yellow stripe beginning near anterior end of snout, running ventral to eye across cheek onto opercle. Iris grayish green dorsally, yellow ventrally. Ground color of body red orange anterodorsally, otherwise mostly rosy; anterior one half of body with two yellow stripes (overlain in places with melanistic pigment)—one just ventral to spinous dorsal fin, the other just ventral to lateral line; two shorter broken yellow stripes ventral to above mentioned stripes. Dorsal and caudal fins mostly red orange. Anal fin mostly dull orange to pink.

Pectoral fin not seen clearly. First three pelvic soft rays mostly orange, remainder of fin more pallid.

The following description is based on notes taken by WDA eight days after the capture and preservation of 11 specimens (63–102 mm SL) of *P. martinicensis*, collected off Cape Lookout, North Carolina, at GMBL station 73-65 in May 1973. (The specimens had lost considerable pigment by the time those notes were written.) Head with greenish-yellow subocular stripe running posteriorly from near anterior end of snout to posterior edge of operculum; two yellow-green stripes radiating from posterodorsal aspect of orbit. Iris orange to red orange. Two to several bars of yellow green on anterior part of body reaching from base of dorsal fin to about level of lower pectoral-fin rays; posterior half of body with greenish-yellow pigmentation, but without well-defined bars; dorsolateral part of body suffused with red orange; ventrolaterally and ventrally body pink to pale. Dorsal fin mainly orange to red orange with some greenish-yellow admixture. Anal fin paler than dorsal fin, with more greenish-yellow pigmentation. Caudal fin similar to anal fin, but frequently with more red to orange. Pectoral fin mostly pallid. Pelvic fin pallid proximally, with some orange distally.

Sexuality. Coleman (1981) provided evidence that *Pronotogrammus martinicensis* is a protogynous hermaphrodite. She found through histological examination 55 females ranging from 46.5 to 112 mm SL, 13 transitional individuals from 73 to 94 mm SL, and 54 males from 66 to 132 mm SL. Immature females in her sample ranged from 46.5 to 72 mm SL. Oogenesis and spermatogenesis appear to have occurred simultaneously in the gonad of a transitional individual of 78 mm SL (Coleman, 1981:893, fig. 1b). McBride et al. (2010) in a histological study of gonads from a large sample of specimens (53 females, 33 transitional, 247 males) of this species from the northeastern Gulf of Mexico corroborated Coleman's findings, with one exception. They found no indication of simultaneous hermaphroditism, and concluded that *P. martinicensis* is a monandric protogynous hermaphrodite.

Reproduction. Trawl-caught ripe females (with vitellogenic oocytes) of *P. martinicensis* have been collected in the eastern Gulf of Mexico during February, March, April, and July (Bullock and Smith, 1991:27). McBride et al. (2010:33) reported females with hydrated oocytes caught in the northeastern Gulf of Mexico during March and May, indicating spring spawning, but females collected in June and August lacked hydrated oocytes, suggesting that "females may not spawn, or may spawn less frequently, during summer. . . ."

Early life history. Baldwin (1990:930–932; figs. 4B, 12, 13) described the early developmental stages of *P. martinicensis* based on 21 specimens, 6.0 to 18.1 mm SL, and included illustrations of four larvae (three of which were taken from Kendall, 1979, as *Anthias*—type 2), and Richards et al. (2006) presented an account of the early life history of the species.

Age and growth. Based on examination of sectioned sagittal otoliths, McBride et al. (2010) estimated ages of 490 specimens of *P. martinicensis* (31–143 mm SL) from the northeastern Gulf of Mexico, as ranging from 0 (i.e., young of the year) to 15 years. Thurman et al. (2004:8; figs. 7, 8) noted differences in growth rates of this species from different study reefs in the northeastern Gulf of Mexico, with mean size-at-age from the several sites varying greatly. Factors producing the disparities in those growth rates are unknown but may be natural (e.g., genetic, availability of food, predation, environmental quality) and/or anthropogenic from varying fishing pressures on larger piscivorous species (Thurman et al., 2004:12).

Maximum length. Smith-Vaniz et al. (1999) reported a specimen (BAMZ 1995-126-024) of *P. martinicensis* from Bermuda (Challenger Bank) that is 163 mm SL—considerably bigger than the longest specimen (132 mm SL) we have seen.

Ecological notes. Parker and Ross (1986:43) made observations of *Pronotogrammus martinicensis* from submersibles off North Carolina. They found this species occurring "singly between 75 and 125 m, often in association with *Oculina* or *Madrepora* clumps." Lindquist and Clavijo (1994) reported on a dive of a submersible on a continental-shelf-edge reef, 95 kilometers southeast of Cape Fear, North Carolina, in depths of 96 to 109 meters, on 8 August 1987; during that dive they observed *P. martinicensis* avoiding light by swimming away in aggregations at the approach of the vessel. Lindquist and Clavijo (1994) estimated the combined population size of *P. martinicensis* and *Baldwinella vivanus* to be 188,000 to 325,000 per 0.5 km^2 in the area explored. Dennis and Bright (1988a:295) remarked that at the Flower Garden Banks in the northwestern Gulf of Mexico, *P. martinicensis* "is most abundant from 80 to 90 m . . . and has been observed down to 150 m" and later mentioned (Dennis and Bright, 1988b:6) that it "is the most conspicuous and common member of the drowned reef community [in the northwestern Gulf], where it was observed to hover up to a meter above drowned reefs from 57 to 140 m." Colin (1974:33) reported *P. martinicensis* as being very common off Jamaica and Belize on rocky outcrops in depths of about 150 to 200 meters. From submersibles, he occasionally observed schools of 10 to 20, but usually saw groups of only two to five fish. Smith-Vaniz et al. (1999:217) noted that small groups of fish believed to be *P. martinicensis* were seen from submersibles on three occasions off Bermuda: "off Castle Roads (160 m), hovering above the substrate, 18 Sep 1983; Challenger Bank (85–105 m), on nearly vertical walls, 27 Feb 1997; Hungry Bay (203 m), on rocky slope, 3 Mar 1997."

Analyses of videotapes made from a ROV and study of gut contents led to the characterization of this species as "a diurnally active mesoplanktivore associated mostly with the base and crest of high profile reefs" in the northeastern Gulf of Mexico (McBride et al., 2010:36, based on Weaver et al., 2002). Bullock and Smith (1991:27) found pteropods, ostracods, and copepods in the stomachs of *P. martinicensis* and noted that "individuals of this species have been regurgitated by *Epinephelus flavolimbatus* and possibly *E. niveatus*" and "taken from the stomach *of Seriola rivoliana . . . .*" We examined four specimens (FMNH 48882, 85–97 mm SL) that were removed from the stomach of a *Seriola falcata* (presumably = *S. rivoliana*) caught in 73 meters off Bermuda.

Brazilian Seaperch. Kuiter (2004:17) under the heading "Brazilian Seaperch" reported the nominal species *Anthias duplicendatus* (as *Holanthias duplicendatus*) from the east coast of Brazil. Anderson and Heemstra (1980) considered *Anthias duplicendatus* Miranda Ribeiro, 1903, to be a junior synonym of *Holanthias martinicensis* (= *Pronotogrammus martinicensis*). Kuiter (2004:17) included a fine illustration of a 14-cm specimen photographed off Rio de Janeiro and made the following comments:

Eastern Brazil. Shallow rocky reefs from 15 to at least 40 m depth. Pinkish with irregular broad reddish band below spinous part of dorsal fin, and pale fins. Eye smallish, its diameter less than snout length. Length to 16 cm. *Remarks*: previously treated as a synonym of *Pronotogrammus martinicensis*, but this needs further investigation.

At a meeting of Brazilian ichthyologists in January 2005, Alfredo Carvalho-Filho and Carlos E. L. Ferreira presented a report (unpublished) on *Pronotogrammus duplicendatus*. They regarded *P. duplicendatus* as a valid species, reported it from several localities off southeastern Brazil in depths of 8 to 40 meters, and compared it with *P. martinicensis*, but they gave no meristic data and indicated neither the number nor the size range of the specimens examined. Carvalho-Filho donated two specimens of the form he and Ferreira called *duplicendatus* to the GMBL collections. The first author carefully examined those specimens (GMBL 06-001, 99 & 127 mm SL, collected off Cabo Frio, ca. 120 kilometers east of Rio de Janeiro, in 25 to 35 meters of water) and found that other than some differences in coloration and the possession of a slightly smaller eye there is nothing that obviously distinguishes them from *P. martinicensis*. Although collected in early January 2006, the specimens from Brazil still had some evidence of live coloration when examined in mid-March 2006, the most distinctive being the presence of magenta (particularly on head, predorsal area, and dorsal fin), pigmentation that we have not seen previously on specimens of *P. martinicensis*. The horizontal diameter of the orbit measured 8.2 and 8.3% SL in the two specimens donated by Carvalho-Filho and 8.6 to 16% SL in 73 other specimens (16–128 mm SL) of *P. martinicensis* examined by us (in 36 specimens, 85–128 mm SL, orbit varied from 8.6 to 12% SL).

More recently, Carvalho-Filho et al. (2009:29) revisited the problem and concluded, based on genetic, morphological, and ecological data, that their *duplicidentatus* "from the Brazilian South-Eastern coast" is "a unique shallow-water variation of" *Pronotogrammus martinicensis*. We agree with that conclusion.

Distribution. Anderson and Heemstra (1980:83) wrote that this species "is known to occur in the western Atlantic from North Carolina to southern Brazil (to as far south as lat. 31.4°S)—including Bermuda, West Indies, Gulf of Mexico and Caribbean Sea." We have examined specimens collected off Bermuda, North Carolina, South Carolina, Atlantic and Gulf coasts of Florida, the Bahamas, in the Straits of Florida, off Alabama, Jamaica, Nicaragua (Caribbean coast), Martinique, Saint Lucia, Venezuela (Caribbean coast), and Brazil (both near the equator and from as far south as latitude 31°22'S). The bottom-associated individuals that we studied were obtained in depths of 25 (25/35) to 229 (155/229) meters.

In addition to localities mentioned by Anderson and Heemstra (1980), Coleman (1982) examined specimens from both the eastern and western Gulf of Mexico, the Caribbean coasts of the Dominican Republic, Honduras, and Colombia, and the coast of Guyana, collected in depths of 38 to 914 meters. Weaver et al. (2006b:61, table 1) observed nearly 6500 individuals of *P. martinicensis* via ROV and submersible at Miller's Ledge in the Tortugas South Ecological Reserve, Florida Keys National Marine Sanctuary. Dennis and Bright (1988b: table 3) listed a number of observations of *P. martinicensis* made from submersibles over 20 different hard-bottom banks in the northwestern Gulf of Mexico, in depths of 57 to 140 meters. Weaver et al. (2006a:87, Appendix A) observed nearly 5000 individuals of *P. martinicensis* during submersible dives at Alderdice Bank in the northwestern Gulf of Mexico. Claro et al. (2000) mentioned that eight specimens were observed from the submersible Johnson-Sea-Link-II at a depth of 196 meters and one at a depth of 177 meters off the southern coast of Cuba. Cervigón (1991:375) reported four specimens (64–117 mm SL) collected west of Isla Margarita, Venezuela, in 112 to 161 meters.

Material examined. One hundred and fifty-four specimens, 16 to 132 mm SL. **BERMUDA:** AMNH 7308 (holotype of *Anthias louisi*: 80 mm SL), FMNH 48882 (4: 85–97), USNM 170003 (1 of 2: 92), USNM 246687 (2: 16–18). **NORTH CAROLINA:** AMNH 43038 (1: 79), AMNH 73647 (2: 61–80), GMBL 60-15 (5: 86–106), GMBL 61-24 (3: 89–104), GMBL 73-65 (10: 63–102), UF 15641 (1: 100), UF 15642 (3: 78–106), UF 230342 (1: 116), UF 230343 (1: 112). **SOUTH CAROLINA:** AMNH 77292 (5: 96–124), GMBL 64-61 (2: 84–104), GMBL 77-197 (2: 74–109), GMBL 81-149 (8: 94–132), GMBL 81-150 (2: 86–112), GMBL 82-156 (1: 71). **FLORIDA (ATLANTIC):** GMBL 61-23 (22: 57–132). **BAHAMAS:** GMBL 61-25 (1: 72), GMBL 61-26 (3: 48–71), GMBL 61-27 (2: 47–71). **STRAITS OF FLORIDA:** GMBL 60-16 (4: 48–85). **FLORIDA (GULF OF MEXICO):** FSBC 11714 (1: 89). **ALABAMA:** GMBL 00-32 (5: 85–97). **JAMAICA:** UF 230505 (1:93). **NICARAGUA (CARIBBEAN):** FMNH 70650 (18: 49–110). **MARTINIQUE:** MNHN 4319 (lectotype of *Aylopon martinicensis*: 97), MNHN B 2767 (3 paralectotypes of *Aylopon martinicensis*: 65–84). **SAINT LUCIA:** ANSP 136024 (1: 84). **VENEZUELA (CARIBBEAN):** GMBL 64-92 (1: 81), UF 228730 (3: 53–68). **BRAZIL (NEAR EQUATOR):** SU 51136 (7: 63–96). **BRAZIL (SOUTHERN):** GMBL 06-001 (2: 99–127), ISH 1824/68 (2: 57–122), ISH 1905/68 (1: 66), MZUSP 11768–11788 (21: 50–128).

Pronotogrammus multifasciatus Gill, 1863

Threadfin Bass

Fig. 28; Tables 2–7; Map 11

Pronotogrammus multifasciatus Gill, 1863:81 (original description; holotype USNM 2762; type locality Cabo San Lucas, Baja California Sur, México).—Fitch, 1982:2 (species account).—Heemstra, 1995:1612 (species account).

Anthias gordensis Wade, 1946:225, pl. 32 (original description; illustration; holotype LACM 21714, 128 mm SL; type locality Inner Gorda Bank, Cabo San Lucas, Baja California Sur, México).

Holanthias sechurae Barton, 1947:2, fig. 2 (original description; illustration; holotype AMNH 17082, 184 mm SL; type locality off Talara, Perú).

Anthias sechurae (Barton): Hildebrand and Barton, 1949:12, fig. 4 (new combination; description; illustration).

Pacificogramma stepanenkoi Kharin, 1983b:117 (original description; illustration; holotype ZIN 45987, 174 mm SL; type locality Uncle Sam Bank, 24°54'N, 112°34'W, Baja California Sur, México).—Anderson and Rosenblatt, 1989:124 (relegated *Pacificogramma stepanenkoi* Kharin, 1983b, to the synonymy of *Pronotogrammus multifasciatus* Gill, 1863).

Diagnosis. *Pronotogrammus multifasciatus* can be distinguished from the only other species of *Pronotogrammus*, *P. martinicensis*, by the following combination of characters. Lateral-line scales 46 to 57 (usually 48–54). Pectoral-fin rays 19 to 21 (usually 19 or 20). Circum-caudal-peduncular scales 23 to 28. Longest dorsal-fin spine (most frequently the fourth) 11 to >17% SL (11–13% SL in specimens more than ca. 170 mm SL).

Description. Dorsal-fin rays X, 15 (rarely X, 14 or XI, 14). Anal-fin rays III, 7 (rarely 8). Gillrakers 10 to 12 + 25 to 30—total 36 to 41 (usually 39–41). Procurrent caudal-fin rays 11 or 12 dorsally and ventrally. Epineurals associated with first 10 or 11 vertebrae.

Vomerine tooth patch usually with posterior prolongation, tooth patch sometimes diamond shaped. Endopterygoids toothless. Tongue usually with teeth, patch(es) of lingual teeth variously shaped. Internarial distance 5 to 15 (usually 6–10) times in snout length. Head length 31 to 36% SL. Snout length 6 to 9% SL. Orbit diameter 7 to 11% SL (7–9% SL in specimens more than ca. 165 mm SL). Body depth at first dorsal spine 27 to 36% SL. Depressed anal-fin length 31 to 35% SL. Pelvic-fin length 25 to 48% SL (30–48% SL in specimens more than ca. 120 mm SL). Caudal-fin lunate to moderately forked, lobes relatively short. Upper caudal-fin lobe 24 to 42% SL (24 to 32% SL in specimens more than ca. 130 mm SL). Lower caudal-fin lobe 22 to 32% SL. Fitch (1982) and Heemstra (1995) presented accounts of this species.

Coloration. In the original description of *P. multifasciatus*, Gill (1863:81) reported that the "color is tawny yellow, with numerous . . . rufous bands descending nearly to the middle, and rather wider than the tawny intervals." Wade (1946:227) in his description of *Anthias gordensis* (= *P. multifasciatus*) noted:

> Body color in alcohol light olive buff. Sides of body dorsally to midline of sides with the scale pattern irregularly outlined in a cross-hatched pattern of dark olive buff. Head body color, shaded slightly darker on opercle. An irregular, dark band extends anteriorly from front edge of eye to symphysis of lower jaw. All fins pale.

Barton (1947:3) mentioned that a 192-mm TL specimen of *Holanthias sechurae* (= *P. multifasciatus*) (presumably in alcohol):

> is yellowish with brown mottling above, a little paler below, fins all pale. It shows traces of three radiating dark lines behind the eye, each line about as wide as pupil, a dark patch between the eyes extending onto snout, and a dark line in front of the eye carried forward onto tip of lower jaw.

Based on a 182-mm SL freshly dead specimen caught off Pt. Fermin, California (33°42'N, 118°20'W), Jones et al. (1985:116) provided the following on coloration: "Dorsal and lateral trunk brilliant red with dark vertical crosshatching. Chin, ventral trunk, caudal peduncle and fin yellow. Head red with dark spotting. Gold stripes radiate from eye to posterior margin of operculum." In addition, Jones et al. (1985:116) commented: "Color in alcohol tan with pale venter. Eye stripes faint." Bussing and López (1994:96) wrote: "Color dusky yellow." Grove and Lavenberg (1997:338) gave the coloration as "Red-pink with yellow blotches, which form irregular stripes; tail and fin membranes yellow." Bussing and López (2005:78) noted: "Head and body rose color, yellow stripe below eye."

Synonymy. We have examined the holotypes of *Pronotogrammus multifasciatus* Gill, 1863, *Anthias gordensis* Wade, 1946, and *Holanthias sechurae* Barton, 1947, and agree with Fitch's (1982) conclusion that *A. gordensis* and *H. sechurae* are conspecific. The similarity of *H. sechurae* to *A. gordensis* was pointed out by J.T.N. (= John Treadwell Nichols) in a footnote in Barton's 1947 paper and was discussed by Hildebrand and Barton (1949) under the new combination *Anthias sechurae*.

Fitch (1982) considered both *A. gordensis* Wade and *H. sechurae* Barton to be junior synonyms of *P. multifasciatus* Gill. But the situation with the name *Pronotogrammus multifasciatus* is extremely problematical. The first author examined the holotype (USNM 2762) of that nominal species in November 1964 and found that the specimen lacked the head and all evidence of fins except a single anal spine. Fitch (1982) reported that a radiograph of the fragment (ca. 30 mm long) of the holotype revealed 16 caudal vertebrae, the number present in all Atlantic and eastern Pacific Anthiinae, except *Anatolanthias apiomycter* which has 15 and *Hypoplectrodes semicinctum* with 17. Assuming, as did Fitch (1982), that Gill's (1863) description is of an anthiine, the only character given by Gill that is of any value in identification to species is the lateral-line scale count of 45. The only anthiine known from the vicinity of Cabo San Lucas, Baja California Sur (the type locality of Gill's species), having a lateral-line scale count of 45 is the species herein referred to as *P. multifasciatus*; our counts of lateral-line scales for this species range from 46 to 57, but Fitch (1982) gave a range of 45 to 51. In view of the poor condition of the type and the limited utility of the original description, it could be argued reasonably that the name *P. multifasciatus* Gill, 1863, should be considered as unidentifiable. In the interest of nomenclatural stability, we choose to follow Fitch (1982) and use the binomen *Pronotogrammus multifasciatus* for this species.

Early life history. Baldwin (1990:932–934; plate 1B; fig. 14) described the early developmental stages of *P. multifasciatus* based on eight specimens, 8.0 to 34.0 mm SL, and included illustrations of four larvae (two of which were taken from Kendall, 1979, as *Anthias gordensis*). Watson (1996:894–895, fig. 8) furnished meristic, morphometric, and life history data for this species and presented illustrations of three larvae (two of which were taken from Kendall, 1979, as *Anthias gordensis*). Beltrán-León and Ríos Herrera (2000) described and illustrated (fig. 124) a larval specimen (5.4 mm) that they identified as *Pronotogrammus* sp. Based on pigmentation and head spination this "larva is almost certainly *P. multifasciatus*" (C. C. Baldwin, *in litt.* to WDA, 7 April 2005).

Ecological notes. Hobson (1975) reported that the stomach of a 102-mm SL specimen of *P. multifasciatus* (as *Anthias gordensis*) collected at Ship Rock, Santa Catalina Island, California (33°28'N, 118°29'W), contained calanoid and cyclopoid copepods, euphausiid larvae, and fish eggs, with calanoids making up 75% of the volume of the identifiable contents and being far more numerous than the combined total number of the other taxa present. Jones et al. (1985) found that the gut contents of a 182-mm SL specimen caught off Pt. Fermin, California, consisted almost entirely of copepods.

Fitch (1982:6) examined specimens of *P. multifasciatus* obtained from the stomachs of two species of fishes—*Seriola lalandi* (Carangidae) and *Epinephelus analogus* (Serranidae), and mentioned that otoliths of *P. multifasciatus* "commonly are found in scats of sea lions, *Zalophus californianus*, that haul out on Islotes Island (north of La Paz), Gulf of California. . . ."

Distribution. We have examined specimens collected off the Pacific coast of the United States (California—nearly to as far north as 34°N), México (Baja California—Pacific and Gulf of California coasts), Costa Rica (Pacific coast and Cocos Island), Panama (Pacific), Colombia (Malpelo Island), Ecuador (Galápagos Islands), and Perú in depths of 42 to 215 (184–215) meters. It has been reported from the Pacific coast of mainland Costa Rica (Bussing and López, 1994, 2005) and off Cocos Island (Bussing and López, 2005). McCosker (*in litt.* to WDA, 6 & 7 March 2007) indicated that this species is extremely common at Cocos Island. The range has been given as Portuguese Bend, Los Angeles County, California (34°N), to the region off Talara, Perú (4°S), and the Galápagos Islands, in depths of 40 to 205 meters (Fitch, 1982; Grove and Lavenberg, 1997). McCosker et al. (1997:24) stated that *P. multifasciatus* "is a widespread species in the eastern Pacific . . . from the outer coast and Gulf of California to Perú" and has been observed in the Galápagos from the research submersible Johnson-Sea-Link and found "to be very abundant along rock reefs to depths of 150–300 m."

Anderson and Rosenblatt (1989:124–125) listed ten records of P. *multifasciatus* from the Southern California Bight, prompting them to suggest that it "should be considered a rare California species . . . rather than an occasional stray."

Material examined. Sixty-six specimens, 48 to 208 mm SL. **CALIFORNIA:** CAS 56855 (1 specimen: 164 mm SL), CAS 56964 (1: 166), LACM 42987-1 (1: 188), LACM 44849-1 (1: 176), LACM 45902-1 (1: 131). **MÉXICO, BAJA CALIFORNIA NORTE (PACIFIC):** SIO 63-161-35A (1: 203); **BAJA CALIFORNIA SUR (PACIFIC):** LACM 30318-1 (2: 144–150), LACM 32630-1 (1: 152), LACM 36161-2 (2: 198–208), LACM 38671-1 (2: 155–173), LACM 42178-5 (1: 151), LACM 49173-1(1: 74); **BAJA CALIFORNIA SUR (CABO SAN LUCAS):** USNM 2762 (holotype of *Pronotogrammus multifasciatus*); **BAJA CALIFORNIA SUR (GULF OF CALIFORNIA, INNER GORDA BANK):** LACM 21713 (paratype of *Anthias gordensis*: 124), LACM 21714 (holotype of *Anthias gordensis*: 128). **COSTA RICA (PACIFIC):** GMBL 73-350 (1: 75), GMBL 74-274 (10: 90–128). **COSTA RICA (COCOS ISLAND):** LACM 32263-6 (1: 176), LACM 32265-2 (2: 128–142), MCZ 49120 (1: 84). **PANAMA (PACIFIC):** MCZ 28774 (4: 77–127). **COLOMBIA (MALPELO ISLAND):** MCZ 28773 (1: 130). **ECUADOR (GALÁPAGOS ISLANDS):** CAS 86501 (2: 104–160), HBOM 107:08472 (3: 151–162), HBOM 107:08473 (4: 135–153), LACM 49544-1 (15: 48–80), USNM 331720 (2: 122–126). **PERÚ:** AMNH 17082 (holotype of *Holanthias sechurae*: 184), AMNH 17083 (paratype of *Holanthias sechurae*: 152).

BALDWINELLA, NEW GENUS

Table 1

Diagnosis. *Baldwinella* is distinguishable from all other genera of Anthiinae covered in this work by the following combination of characters. Scales ctenoid, with only marginal cteni, no ctenial bases in posterior fields (Fig. 1B). Maxilla without scales. Gillrakers well developed, total number on first arch 36 to 43. Lateral-line scales 36 to 53 (usually fewer than 51). Circum-caudal-peduncular scales 22 to 29. Preopercle serrate but without antrorse spines on ventral border. Urohyal without anteriorly projecting spine.

Baldwin (1990:918–921; figs. 1B, 2, 4A; pl. 1A) reported a number of larval characters that are apparently synapomorphies for the species of this genus. They are: frontal and parietal with serrate ridges, supraoccipital with serrate "cockscomb" crest, serrate horizontal ridge on pterotic, articular with serrate ventral margin, pelvic fin with serrate spine, and type A larval scales.

Description. Premaxillae protrusile. No supramaxilla. Anterior and posterior nares rather closely set on each side of snout; posterior border of anterior naris never produced into a long filament. No fleshy papillae on border of orbit. Outer teeth in jaws mostly conical; inner teeth mostly villiform or cardiform; some enlarged as canines. Vomerine tooth patch chevron shaped (patch rarely triangular), without posterior prolongation. Band of teeth on each palatine. No teeth on endopterygoids or tongue.

Single dorsal fin; spinous part of fin incised at junction of spinous and soft-rayed parts; dorsal-fin rays X, 13 to 16 (usually X, 14 or 15; very rarely with XI spines). Anal-fin rays III, 7 to 9 (usually 8). Pectoral fin roughly symmetrical with 15 to 21 (usually 16–19) rays. Principal caudal-fin rays 15 (8 + 7); branched rays 13 (7 + 6); procurrent rays 8 to 14 dorsally and ventrally. Vertebrae 26 (10 + 16). First caudal vertebra without parapophyses. Formula for configuration of supraneural bones, etc. 0/0/2/1+1/1/ (Fig. 2B). Pleural ribs on vertebrae 3 through 10. Epineurals associated with first 10 to 13 vertebrae. No trisegmental pterygiophores associated with dorsal and anal fins.

Lateral line complete, extending to at least base of caudal fin; at midbody, lateral line running parallel to dorsal body contour a few scale rows below dorsal fin, then descending rather steeply to continue near midlateral axis of caudal peduncle. No secondary squamation. Much of head covered with scales, but snout, lachrymal region, maxilla, most of interorbital, gular region, branchiostegals, and branchiostegal membranes without scales; lower jaw with or without scales. Dorsal and anal fins mostly without scales, although soft parts of those fins may be more or less scaly basally; pectoral, pelvic, and caudal fins scaly basally. Modified scales (interpelvic process) overlapping pelvic-fin bases along midventral line.

Etymology. The name *Baldwinella* (gender feminine) is for Carole C. Baldwin (Division of Fishes, National Museum of Natural History, Smithsonian Institution, Washington, D.C.) in recognition of her contributions to the understanding of the systematics of serranid fishes.

Type species. *Pronotogrammus aureorubens* Longley, 1935.

Key to the Species of *Baldwinella*

1a.   Lateral-line scales 36–40; posterior half or more of lower jaw scaly . . . .
. . . . *Baldwinella eos*
(eastern North Pacific)

1b.   Lateral-line scales 42–53; lower jaw with or without scales 2

2a.   Posterior 30–60% of lower jaw scaly; dorsal-fin spines without long filaments (spines usually with short filamentous tabs); preopercle serrate, but angle of preopercle without well-developed spine in specimens more than about 45 mm SL; orbit diameter 11–19% SL; pectoral-fin rays 15–19 (usually 16 or 17); soft rays in dorsal fin 13–16 (usually 15, rarely 13 or 16) . . . . *Baldwinella aureorubens*
(western North Atlantic)

2b.   Lower jaw without scales; usually, several dorsal-fin spines with long filaments, filament of fourth dorsal-fin spine up to 74% SL (Thurman et al., 2004); preopercle serrate, usually with single spine or spine-like process at angle, angle sometimes with bifid spine or multiple spines; orbit diameter 7–11% SL; pectoral-fin rays 16–21 (usually 18 or 19); soft rays in dorsal fin 13 or 14 (usually 14) . . . . *Baldwinella vivanus*
(western Atlantic)

Baldwinella aureorubens (Longley, 1935)

Streamer Bass

Fig. 29; Tables 2–7; Map 12

Pronotogrammus aureorubens Longley, 1935:88 (original description; holotype USNM 101320, 169 mm SL; type locality south of Tortugas, Florida).—Longley and Hildebrand, 1940:242, fig. 10 (description, illustration).

Hemanthias aureorubens: combination used by many authors, including Anderson, 2003:1350 (species account).

Diagnosis. A species of *Baldwinella* distinguishable from the other members of the genus by the following combination of characters. Tubed scales in the lateral line 42 to 50. Posterior 30 to 60% of lower jaw scaly. Membranes of dorsal fin produced into short filaments at tips of spines, but never produced to the extent seen in filaments of spines in middle of spinous dorsal fin of *Baldwinella vivanus*. Preopercle serrate, but angle of preopercle without well-developed spine in specimens more than about 45 mm SL. Orbit diameter 11 to 19% SL. Pectoral-fin rays 15 to 19 (usually 16 or 17). Soft rays in dorsal fin 13 to 16 (usually 15, rarely 13 or 16).

Description. Dorsal-fin rays X or XI, 13 to 16 (usually X, 15; rarely with XI spines). Anal-fin rays III, 7 to 9 (usually 8). Gillrakers 10 to 13 + 26 to 31—total 36 to 43 (usually 38–41). Lateral-line scales 42 to 50 (usually 43–48). Circum-caudal-peduncular scales 22 to 27 (usually 23–26). Procurrent caudal-fin rays 9 or 10 dorsally and ventrally. Epineurals associated with first 11 to 13 vertebrae.

Internarial distance 7 to 17 times in snout length. Head length 31 to 41% SL (34–38% SL in specimens more than about 170 mm SL). Snout length 6 to 11% SL. Orbit diameter 11 to 19% SL (11 to 16% SL in specimens more than about 90 mm SL). Body depth at first dorsal-fin spine 31 to 40% SL. Longest dorsal-fin spine (third, fourth, fifth, sixth, or seventh—usually the fourth), 10 to 21% SL (10–15% SL in specimens more than ca. 80 mm SL). Depressed anal-fin length 26 to 36% SL (26–30% SL in specimens more than ca. 160 mm SL). Pelvic-fin length 24 to 36% SL (24 to 29% SL in specimens greater than about 170 mm SL). Caudal fin deeply forked; upper lobe almost always longer than lower; upper lobe 40 to 126% SL; lower lobe 37 to 116% SL. Anderson (2003) gave a short species account.

Coloration. Bullock and Smith (1991:207, pl. I, fig. C) wrote (p. 18): "Color pinkish dorsally; scales yellow-margined; silvery belly and sides. Dorsal and caudal fins yellow; pectorals pink; other fins pale." Matsuura (1983:313) published a color photograph of this species (as *Pronotogrammus eos*) and described the coloration as: "Body pale red dorsally, silver-white ventrally; fins pale yellow." A color photograph, taken by Donald D. Flescher, of a specimen of *Baldwinella aureorubens* of 197 mm SL collected off New Jersey (GMBL 82-30) shows the following: Dorsum of head and body rosy; lateral and ventral aspects of head and body mostly silvery. Iris with inner circle of yellow and an outer circle of rosy to grayish. Dorsal fin yellowish along spines and on distal parts of soft rays. Pectoral fin rosy. Anal and pelvic fins pallid. Outermost elements of upper and lower caudal-fin lobes dull yellow to yellow green; middle of caudal fin pallid.

Reproduction. Bullock and Smith (1991) found vitellogenic oocytes in females (116–151 mm SL) collected during May 1984 off the east coast of Florida.

Early life history. Baldwin (1990) provided a description of the early developmental stages of *B. aureorubens* (cited as *Pronotogrammus aureorubens*) based on six specimens, 10.8 to 26.0 mm SL, and included illustrations (fig. 7) of two larvae (one of which was taken from Kendall, 1984), and Richards et al. (2006) presented an account of the early life history of the species (as *Pronotogrammus aureorubens*).

Ecological note. Specimens of *Baldwinella aureorubens* have been found in the stomach contents of *Pagrus pagrus* and *Haemulon melanurum* collected on the Flower Garden Banks off Texas (Bullock and Smith, 1991, citing Nelson, 1988).

Distribution. We have examined specimens of *Baldwinella aureorubens* collected off the Atlantic coast of the United States (New Jersey, Georgia, Florida), in the Straits of Florida, off Dry Tortugas (Florida), off the Gulf coast of the United States (Florida, Louisiana, Texas), off the Caribbean coast of South America (Colombia, Venezuela), and off the Atlantic coast of South America (Venezuela, Suriname, French Guiana) in depths of 146 (146/165) to 439 meters.

A juvenile *B. aureorubens* (MCZ 85706, 29 mm SL) was taken in September of 1962 well off the coast of Massachusetts by midwater trawl, fished mainly at a depth of 64 meters (Karsten Hartel, *in litt.* to WDA, 14 July 2000). Hartel provided a photograph of another juvenile (MCZ 164522, 30 mm SL) that was collected off Bear Seamount (@39°57'N, 67°30'W; depth 0–1428 meters; by midwater trawl fished open and held at maximum depth for some time; 05 June 2004; Hartel, *in litt.* to WDA, 23 February & 15 March 2005). As suggested by Hartel (*in litt.* to WDA, 23 February 2005), small juveniles such as those MCZ specimens are probably near the size at which settling out of the plankton begins. Moore et al. (2003:225) reported adult *B. aureorubens* from four localities off northeastern United States.

Longley and Hildebrand (1940:242) observed that this species "is rather common south of Tortugas at depths of 100 to 200 fathoms." Houde (1982) reported the collection of larvae of *B. aureorubens* (as *Pronotogrammus aureorubens*) from the eastern Gulf of Mexico. Roa-Varón et al. (2003) recorded this species from the Caribbean coast of Colombia. Cervigón (1966:338), in reporting it from off Venezuela, gave a depth range of "50 a 250 brazas" (91 to 457 meters) and in a subsequent publication (Cervigón, 1991), approximately 120 to 610 meters. Gines and Cervigón (1968) included *B. aureorubens* (as *Pronotogrammus aureorubens*) in a list of fishes collected from the coasts of Guyana and Suriname in 1967, stating (p. 32) that it is one of the species typical of depths between 100 and 200 fathoms (183 and 366 meters) and citing specifically two localities of capture (pp. 85–86) off Suriname in 136 to 145 fathoms (249 to 265 meters).

Material examined. One hundred and forty specimens, 38 to 212 mm SL. **NEW JERSEY:** GMBL 82-30 (7 specimens: 190–212 mm SL), MCZ 162190 (1: 108), SHML uncat. (1: 200). **GEORGIA:** GMBL 72-30 (1: 157), GMBL 75-122 (1: 157). **FLORIDA (ATLANTIC):** FMNH 70670 (5: 95–188), GMBL 61-29 (2: 120–121), UF 230330 (1: 188), UF 230339 (3: 117–168). **STRAITS OF FLORIDA:** FMNH 70661 (2: 168–200), FMNH 70663 (1: 115), FMNH 70664 (5: 108–174), TU 10900 (5: 93–181), TU 10914 (1: 154), UF 46780 (5: 66–87), UF 216242 (1: 159), USNM 157824 (1: 140). **FLORIDA (GULF OF MEXICO—DRY TORTUGAS):** USNM 101320 (holotype: 169), USNM 117226 (1: 118). **FLORIDA (GULF OF MEXICO):** FMNH 66269 (1: 145), UAIC 8219.02 (1: 49), UF 12292 (2: 144–177), UF 71453 (1: 68), UF 71536 (3: 89–112), UF 72232 (1: 128), UF 72250 (2: 71–91), USNM 185392 (1: 150), USNM 331718 (1: 155), USNM 333310 (2: 89–171). **LOUISIANA:** FMNH 70666 (1: 193), FMNH 70668 (1: 140), GMBL 62-47 (1: 129), USNM 185505 (3: 150–174), USNM 331719 (1: 168). **TEXAS:** FMNH 70665 (2: 137–170), FMNH 70667 (1: 168), TU 2768 (2: 43–141), UF 40246 (4: 105–210). **COLOMBIA (CARIBBEAN):** GMBL 64-69 (1: 145), GMBL 64-70 (1: 73), MCZ 45936 (3: 111–159), UF 217167 (1: 181). **VENEZUELA (CARIBBEAN):** FMNH 70660 (1: 195), FMNH 70662 (2: 145–154), GMBL 60-21 (4: 117–181), UF 215418 (4: 117–149), UF 228727 (1: 171), USNM 188996 (8: 89–137). **VENEZUELA (ATLANTIC):** FMNH 70655 (1: 132), FMNH 70656 (9: 62–162), FMNH 70657 (11: 38–179), USNM 185228 (2 of 7: 83–97), USNM 185478 (1: 167), USNM 185481 (3: 159–166). **SURINAME:** FMNH 70713 (6: 110–143), GMBL 58-13 (1: 120), GMBL 58-14 (1: 84), USNM 185416 (1: 115). **FRENCH GUIANA:** USNM 185423 (2: 94–126).

Baldwinella eos (Gilbert, 1890)

Largescale Jewelfish

Fig. 30; Tables 2–7; Map 11

Pronotogrammus eos Gilbert, 1890:62 (original description; lectotype, herein designated, SU 229, 143 mm SL; type locality Gulf of California [ALBATROSS station 2996: 24°30'15"N, 110°29'00"W; 112 fathoms]).—Fitch, 1982:3, fig. 2 (species account, illustration).—Heemstra, 1995:1611 (species account).

Diagnosis. A species of *Baldwinella* distinguishable from the other members of the genus by the number of tubed scales in the lateral line: 36 to 40 in *B. eos* vs. 42 to 53 in the other two species. Other characters useful in distinguishing this species from

B. vivanus are: Posterior half or more of lower jaw scaly. Membranes of dorsal fin produced into short filaments at tips of spines, but never produced to the extent seen in filaments of spines in middle of spinous dorsal fin of *B. vivanus*. Preopercle serrate, but larger juveniles and adults without well-developed spines or spine-like processes at angle. Orbit diameter 12 to 15% SL.

Description. Dorsal-fin rays X, 14 or 15 (usually 15). Anal-fin rays III, 7 to 9 (almost always 8). Pectoral-fin rays 16 to 19 (usually 17, rarely 16 or 19). Gillrakers 10 to 13 + 27 to 30—total 38 to 43 (usually 39–41). Lateral-line scales 36 to 40. Circum-caudal-peduncular scales 22 to 25. Procurrent caudal-fin rays 8 to 10 dorsally, 8 or 9 ventrally. Epineurals associated with first 11 or 12 vertebrae.

Internarial distance 6 to 13 times in snout length. Head length 37 to 44% SL. Snout length 5 to 9% SL. Body depth at first dorsal-fin spine 32 to 37% SL. Longest dorsal-fin spine (fifth, sixth, or seventh—most frequently the sixth) 13 to 18% SL. Depressed anal-fin length 33 to 38% SL. Pelvic-fin length 29 to 36% SL. Caudal fin lunate to deeply forked; upper lobe usually longer than lower; upper lobe 33 to 49% SL (33–37% SL in specimens more than about 120 mm SL); lower lobe 30 to 46% SL (30–37% SL in specimens more than about 120 mm SL). Fitch (1982) and Heemstra (1995) presented accounts of this species.

Coloration. Gilbert (1890:63–64) described the coloration as:

> Rosy red, overlying silvery on sides, and below the fins light yellow. A dusky spot above the middle of each orbit, and two V-shaped olive-brown marks behind the head, one from nape downward and backward on each side to upper angle of gill openings, the second parallel with it, starting from origin of dorsal. Lining of buccal and gill cavities, and peritoneum silvery white.

Bussing and López (1994:96) wrote: "2 dark bars on nape; color pink. . . ." Heemstra (1995:1611) reported: *"Color: cabeza y cuerpo de tonos plateados, dorso y aletas medianas rosáceos; aletas pélvicas de color amarillo claro."* [= Color: head and body of silvery tones, dorsum and median fins rosy; pelvic fins of light yellow color.] Bussing and López (2005:78) noted: "Head and body red above, silvery below; 2 V-shaped bars crossing nape; yellow-spotted fins."

Designation of lectotype. Gilbert (1890) did not designate a holotype when he described the species. Consequently, the type series included eight syntypes (all from ALBATROSS station 2996 in the Gulf of California): two, SU 229—143 & 149 mm SL; one, USNM 44387—162 mm SL; three, USNM 46550—110 to 114 mm SL; two, USNM 125366—138 & 153 mm SL. To unequivocally fix the name of the species to a zoological entity, we hereby designate as the lectotype of *Pronotogrammus eos* Gilbert, 1890, the smaller (143 mm SL) of the two SU syntypes, which retains SU 229 as its catalogue number. By that action the other SU syntype and the USNM syntypes become paralectotypes; the SU paralectotype has been assigned a new catalog number (SU 69848).

Early life history. Baldwin (1990:925–926; plate 1A; figs. 2, 8) described the early developmental stages of *B. eos* (cited as *Pronotogrammus eos*) based on five specimens, 9.3 to 24.0 mm SL, and included illustrations of two larvae (one of which was taken from Kendall, 1979, as *Hemanthias peruanus*).

Ecological note. Fitch (1982:6) mentioned that otoliths of this species "commonly are found in scats of sea lions, *Zalophus californianus*, that haul out on Islotes Island (north of La Paz), Gulf of California. . . ."

Distribution. *Baldwinella eos* ranges from "mid-Gulf of California (28°N)" (Fitch, 1982:3) to Colombia (6.5°N) and is known from off Cocos Island (Bussing and López, 2005). We examined specimens collected in 82 to 393 (247/393) meters. Fitch (1982) gave the depth range as 115 to 325 meters.

Material examined. One hundred and seventeen specimens, 30 to 162 mm SL. **MEXICO, BAJA CALIFORNIA SUR, NORTH OF LA PAZ (GULF OF CALIFORNIA):** SU 229 (lectotype: 143 mm SL), SU 69848 (1 paralectotype: 149), USNM 44387 (1 paralectotype: 162), USNM 46550 (3 paralectotypes: 110–114), USNM 125366 (2 paralectotypes: 138 & 153). **COSTA RICA (PACIFIC):** GMBL 73-163 (9: 81–111), GMBL 73-189 (5: 95–122), GMBL 73-347 (16: 64–143), GMBL 73-349 (4: 88–156), GMBL 73-352 (1: 52), GMBL 73-353 (9: 46–103), GMBL 73-355 (2: 101 & 117), GMBL 74-273 (1: 102), GMBL 74-275 (4: 81–124), GMBL 74-277 (4: 37–60), GMBL 74-279 (1: 30), GMBL 74-282 (1: 89), GMBL 74-286 (1: 141), GMBL 02-107 (11: 82–116), LACM 33828-2 (1: 50). **PANAMA (PACIFIC):** MCZ 28775 (2: 86 & 141), MCZ 49121 (2: 98 & 122), UF 229714 (18: 89–145), CANOPUS cruise—station 33 (1: 107). **COLOMBIA (PACIFIC):** UF 230478 (6: 105–128), UF 230479 (10: 97–140).

Baldwinella vivanus (Jordan and Swain, 1885)

Barbier Rouge

Fig. 31; Tables 2–7; Map 12

Anthias vivanus Jordan and Swain, 1885:544 (original description; holotype USNM 36942, 57 mm SL; type locality off Pensacola, Florida).

Hemanthias vivanus, combination used by many authors, including Anderson, 2003:1352 (species account).

Diagnosis. A species of *Baldwinella* distinguishable from the other members of the genus by the absence of scales on the lower jaw. The following combination of characters is also useful in distinguishing *B. vivanus* from the other species of *Baldwinella*. Orbit diameter 7 to 11% SL. Usually, several dorsal-fin spines with long filaments, filament of fourth dorsal-fin spine up to 74% SL (Thurman et al., 2004). Preopercle serrate, usually with single spine or spine-like process at angle, angle sometimes with bifid spine or multiple spines. Soft rays in dorsal fin 13 or 14 (usually 14). Pectoral-fin rays 16 to 21 (usually 18 or 19). Lateral-line scales 44 to 53.

Description. Dorsal-fin rays X, 13 or 14 (rarely 13). Anal-fin rays III, 8 or 9 (very rarely 9). Pectoral-fin rays 16 to 21 (18 or 19 in more than 90% of counts). Gillrakers 10 to 13 + 27 to 31—total 38 to 43 (39–42 in more than 90% of counts). Lateral-line scales 44 to 53 (46–51 in 85% of counts). Circum-caudal-peduncular scales 24 to 29 (rarely 24 or 25). Procurrent caudal-fin rays 13 or 14 dorsally and ventrally. Epineurals associated with first 10 to 12 vertebrae.

Internarial distance 5 to 9 times in snout length. Head length 28 to 32% SL. Snout length 6 to 9% SL. Body depth at first dorsal-fin spine 26 to 35% SL (28–32% SL in 88% of specimens examined). Longest dorsal-fin spine (fourth) 13 to 18% SL; fin membranes extending into filaments at distal tips of first nine dorsal spines; filaments, particularly those of spines in middle of spinous dorsal fin, often quite elongated; filament of fourth dorsal spine up to 52% SL in specimens we examined (Hastings, 1981:445, reported it to be up to 65% SL, and Thurman et al., 2004:10, up to 74% SL). Depressed anal-fin length 35 to 45% SL (42–45% SL in specimens more than ca. 115 mm SL). Pelvic-fin length 28 to 38% SL. Caudal fin deeply forked; upper lobe usually longer than lower; upper lobe 40 to >70% SL; lower lobe 31 to 68% SL. Anderson (2003) furnished a short species account.

Coloration. Bullock and Smith (1991:209, pl. II, fig. B) presented a color photograph of *Baldwinella vivanus* of ca. 125 mm and wrote (p. 23): "Body color carmine; olive freckles on back and sides; two gold stripes on head."

A color photograph, taken by Donald D. Flescher, of a specimen of *Baldwinella vivanus* of 109 mm SL collected off North Carolina (GMBL 80–336) shows the following: head reddish dorsally, pinkish ventrally; yellow stripe on snout which splits into two stripes posteriorly—dorsal stripe extending from posterior border of orbit to posterior end of opercle near pectoral-fin base, ventral stripe running somewhat obliquely beneath orbit and continuing to posterior end of head. Iris rather dully colored—anteriorly yellowish, dorsally gray (centrally) to almost black (peripherally), ventrally yellow (centrally) to almost black (peripherally), posteriorly yellow (centrally) to rosy (peripherally). Body rosy with admixture of purple dorsally, white ventrally. Dorsal fin mostly red-orange to rose. Anal fin bright yellow. Pectoral fin not seen clearly. Pelvic fin with leading edge yellow, trailing edge orange to red orange, pale medially. Caudal fin mostly gray. The coloration of the anal fin is like that described for males by Hastings (1981), but the coloration of the pelvic-fin does not resemble Hastings' description for either sex (see **Sexuality, dimorphism, and dichromatism**). Because the filaments of the dorsal spines are well developed (those of dorsal spines four to seven being quite long, that of spine four being ca. 51% SL), this specimen is most likely a male (see **Sexuality, dimorphism, and dichromatism**).

Upon examining a color transparency made by George C. Miller of a specimen of 131 mm SL (GMBL 65-28), PCH made the following notes: "Body generally pinkish; yellow on margin of anal fin; pelvics, tips of caudal fin, elongate dorsal rays, iris and snout also yellow."

Sexuality, dimorphism, and dichromatism. Hastings (1981) demonstrated that *Baldwinella vivanus* is a protogynous hermaphrodite and showed that the species is sexually dimorphic in lengths of filaments of the dorsal-fin spines and sexually dichromatic. The filaments of the dorsal-fin spines are relatively short in females, but become elongated in males (Hastings, 1981; Thurman et al., 2004). Thurman et al. (2004:10; fig. 17) reported the lengths of the filament of the fourth dorsal-fin spine (in % SL) as follows: 16 females—6 to 30 (mean 16), 8 transitional specimens—8 to 52 (mean 29), and 100 males—11 to 74 (mean 49). Hastings (1981:444–445) reported that males and females are similar in coloration "except that the anal fin, which is mottled with blue and olive in females, becomes bright yellow in males. Additionally, the pelvic fins are pink in females and become blood red in males."

Hastings (1981:445, table 1) gave the following ranges in mm SL for specimens collected off the Atlantic coast of Florida: 49 females 49 to 77, 7 transitional individuals 65 to 74, 29 males 65 to 96. In contrast, Bullock and Smith's (1991:24) material from the eastern Gulf of Mexico included larger individuals of both sexes (females 67–117 mm SL, transitional individuals 95–106, males 113–117). In samples from the northeastern Gulf of Mexico, Thurman et al. (2004:10; fig. 15) found smaller males (one each at 46 and 62 mm SL) and all specimens greater than 77 mm SL to be males.

Reproduction. Bullock and Smith (1991:24) reported ripe specimens of *B. vivanus* collected in April through June and in August.

Early life history. Scott and Scott (1988:364) reported several small specimens (3.5 mm TL—25.0 mm SL) of this species (identification verified by C. C. Baldwin) that were taken off Georges Bank; Baldwin (1990:920, 922–924; figs. 1B, 3–6) described the early developmental stages of *B. vivanus* (cited as *Hemanthias vivanus*) based on 197 specimens, 2.0 mm NL to 35.0 mm SL, and included illustrations of five larvae (three of which were taken from Kendall, 1979); and Richards et al. (2006) presented an account of the early life history of the species (as *Hemanthias vivanus*).

Age and growth. Estimates of age of *B. vivanus* from the northeastern Gulf of Mexico range from age-0 to age-VIII (Thurman et al., 2004:9; fig. 13). Growth slowed by age-II (Thurman et al., 2004:10; fig. 14), meaning that most somatic increase in size took place in the first two years of life.

Ecological notes. Bullock and Smith (1991) examined the stomachs of 33 specimens of *B. vivanus* and found food in eight of them—the most important items being calanoid and cyclopoid copepods, amphipods, and ostracods. Lindquist and Clavijo (1994) found food in the guts of 40 of 51 preserved specimens (30–77 mm SL, mean 43 mm) collected with quinaldine off Cape Fear, North Carolina; food items identified were harpacticoid, cyclopoid, and calanoid copepods and foraminifera—with harpacticoids making up 94.3% by volume of the gut contents. Lindquist and Clavijo (1994:139) opined that their feeding data for *B. vivanus* suggest that "juveniles and small females . . . are benthic or near-benthic water column foragers on harpacticoid copepods." Because the gut contents examined came from specimens collected at midday and the prey items found were in the posterior portions of the guts, suggest "an early morning feeding period," but since it is not known "if the fish regurgitated upon ascent or as a result of quinaldine narcosis, we cannot be conclusive about the feeding period" (Lindquist and Clavijo, 1994:139).

Fishes known to prey upon *B. vivanus* include the groupers *Epinephelus flavolimbatus*, *E. niveatus*, and *E. drummondhayi* (Bullock and Smith, 1991). Jordan and Evermann (1896:1224) stated: "nearly all of them [specimens of *Baldwinella vivanus* then known] being from the spewings of the speckled Hind, *Epinephelus drummondhayi*." One of the specimens of *B. vivanus* (GMBL 77-291, 69 mm SL) examined by us is from the stomach of an *E. drummondhayi* caught off Frying Pan Shoals, North Carolina, and another (FSBC 11649, 101 mm SL) was regurgitated by an *E. flavolimbatus* taken in the Gulf of Mexico off the coast of Florida.

Jordan and Swain (1885:544) reported that the holotype of *B. vivanus* "was taken from the stomach of a red snapper (*Lutjanus vivanus*) at Pensacola . . . ." If taken from the stomach of a red snapper, the predator was a specimen of *L. campechanus*. That presumption is supported by the comment on the derivation of the specific name *vivanus* by Jordan and Evermann (1896:1224), i.e., "From the Red Snapper or *Vivanet*, then called *Lutjanus vivanus*, from the stomach of which this species was first taken."

Parker and Ross (1986) observed *B. vivanus* from submersibles on 9 of 10 reefs examined in depths between 52 and 152 meters off North Carolina. They noted (p. 43) that this species "usually appeared in large, fast-moving schools"; most of the members of those schools appeared to be juveniles less than 150 mm TL. In one instance they saw *B. vivanus* use a burrow along with two other serranids (*Diplectrum formosum* and *Liopropoma eukrines*).

Lindquist and Clavijo (1994) reported on a dive of a submersible on a continental-shelf-edge reef, 95 kilometers southeast of Cape Fear, North Carolina, in depths of 96 to 109 meters, 8 August 1987; during that dive they observed *B. vivanus* avoiding light by darting into crevices and holes. They estimated the combined population size of *B. vivanus* and *Pronotogrammus martinicensis* to be 188,000 to 325,000 per 0.5 km^2 in the area explored. Gutherz et al. (1995: table 2) observed, from a submersible operating off Charleston in depths of 185 to 220 meters, small numbers of *B. vivanus*.

During a submersible dive off Cape Fear, North Carolina, 58 specimens of *B. vivanus* were collected in 96 meters; the 27 fish that survived the stress of capture were transferred to a 50-liter container. Over the next 18 hours, 24 of the survivors died. The remaining three fish lived for five months in a 114-liter aquarium and "appeared to behave and feed normally (a size related dominance hierarchy was noted) and were healthy until they died from a fungal or bacterial infection" (Lindquist and Clavijo, 1994:136–138).

Distribution. We have examined specimens collected off the Atlantic coast of the United States (Delaware, North Carolina, South Carolina, Georgia, Florida), in the Bahamas, in the Straits of Florida, off the Gulf coast of the United States (Florida, Alabama, Louisiana), in the Caribbean Sea (off Belize, Nicaragua, Colombia, Venezuela), and off the Atlantic coast of South America (Venezuela, southern Brazil— latitude 33°36'S) in depths of 22 to 274 meters.

Scott and Scott (1988:364) reported larvae of this species caught off Georges Bank at ca. 41°21'N, 66°14'W (see **Early life history**). Moore et al. (2003:225) reported *B. vivanus* from a locality east of Rehoboth Beach, Delaware, near Wilmington Canyon in 0 to 128 meters. Weaver et al. (2006b:61, table 1) observed more than 3300 individuals of *B. vivanus* via ROV and submersible at Miller's Ledge in the Tortugas South Ecological Reserve, Florida Keys National Marine Sanctuary. Bullock and Smith (1991) examined specimens collected south of Dry Tortugas and from the eastern Gulf of Mexico in depths of 73 to 427 meters, and Houde (1982) noted the collection of larvae from the eastern Gulf of Mexico. Cervigón (1991:370) reported two specimens (65 and 95 mm SL) collected west of Isla Margarita, Venezuela, in depths of 112 to 161 meters. Schaldach et al. (1997) included *B. vivanus* (as *Hemanthias vivanus*) in a list of marine fishes known from the region of Los Tuxtlas (State of Veracruz, México).

Material examined. One hundred and ninety-nine specimens, 33 to 143 mm SL. **DELAWARE:** MCZ 162089 (3 specimens: 112–143 mm SL), MCZ 162090 (1: 115). **NORTH CAROLINA:** ANSP 127494 (2: 74–89), FMNH 66005 (6: 66–100), FMNH 70700 (2: 92–98), GMBL 59-32 (3: 78–98), GMBL 59-44 (1: 88), GMBL

60-27 (30: 45–94), GMBL 64-88 (4: 44–97), GMBL 64-89 (2: 90–96), GMBL 64-90 (3: 61–64), GMBL 77-291 (1: 69), GMBL 80-336 (1: 109), UF 15601 (3: 73–91), UF 40764 (1: 81), UF 40798 (2: 62–71), UF 41837 (1: 57), UF 230344 (1: 87), USNM 134182 (2: 92–113), USNM 190340 (2: 65–74). **SOUTH CAROLINA:** AMNH 77308(3: 86–132), GMBL 60-13 (1: 88), GMBL 75-105 (1: 60), GMBL 82-29 (4: 92–141), GMBL 82-156 (6: 56–97). **GEORGIA:** GMBL 63-85 (1: 33), GMBL 64-59 (1: 68). **FLORIDA (ATLANTIC):** GMBL 61-37 (8: 57–106), GMBL 64-91 (1: 51), TU 14798 (1: 94), UF 202669 (2: 81–102). **BAHAMAS:** GMBL 60-14 (6: 55–74), GMBL 65-28 (1: 131). **STRAITS OF FLORIDA:** GMBL 63-86 (2: 45–77), UF 218530 (1: 62). **FLORIDA (GULF OF MEXICO):** FMNH 45478 (3: 104–117), FMNH 57014 (1: 90), FMNH 61335 (10: 61–93), FMNH 66004 (1: 75), FMNH 70698 (7: 83–106), FMNH 70699 (3: 60–92), FMNH 70701 (7: 102–129), FSBC 11649 (1: 101), GMBL 62-59 (2: 81–98); MCZ 27129 (2: 83–87), SAIAB 44590 (1: 113), UAIC 8026.02 (4: 76–106), UAIC 8027.02 (1: 90), UAIC 8035.04 (7: 43–114), UAIC 15275.01 (2: 105–116), USNM 36942 (holotype of *Anthias vivanus*: 57), USNM 328276 (1 of 2: 90), USNM 328277 (1: 80). **ALABAMA:** GMBL 00-32 (6: 67–99), UF 228651 (10: 88–105), USNM 158259 (8: 79–102). **LOUISIANA:** UF 217168 (1: 82). **BELIZE:** FMNH 70694 (1: 98). **NICARAGUA (CARIBBEAN):** FMNH 70695 (2: 92–103), FMNH 70696 (1: 83), FMNH 70697 (1: 84). **COLOMBIA (CARIBBEAN):** FMNH 70692 (1: 73). **VENEZUELA (CARIBBEAN):** GMBL 64–60 (1: 108). **VENEZUELA (ATLANTIC):** USNM 185157 (1: 103). **SOUTHERN BRAZIL:** MZUSP 11763 (1: 100).

ANATOLANTHIAS ANDERSON, PARIN, AND RANDALL, 1990

Table 1

Anatolanthias Anderson, Parin, and Randall, 1990:923 (type species *Anatolanthias apiomycter* Anderson, Parin, and Randall, 1990:924, by original designation).

Diagnosis. *Anatolanthias* is easily distinguished from the other genera of Anthiinae covered herein by the following characters. Anterior naris located rather far anteriorly on snout, somewhat remote from posterior naris; internarial distance 2.8 to 3.1 times in snout length. Fleshy papillae present on posterior half of orbital border. Vomer edentate. (Species of other genera with nares closer together, internarial distance 3–30, usually 6–17, times in snout length; without orbital papillae; and with vomerine dentition). Additional characters useful in combination for distinguishing *Anatolanthias* from the other genera described herein are: Scales ctenoid, with only marginal cteni, no ctenial bases in posterior fields (Fig. 1B). Maxilla scaly. Lateral-line scales 62 or 63.

Description. Premaxillae protrusile. No supramaxilla. Maxilla abruptly expanded distally, particularly on labial border, where a shelf or rostrally directed hook is present at point of expansion. Posterior border of anterior naris produced into flap that falls well short of posterior naris when reflected. Posterior margin of preopercle serrate; ventral margin of preopercle essentially smooth, without antrorse spines. Teeth in jaws mostly conical; a few canines anteriorly in each jaw. Vomer, endopterygoids, and tongue edentate. Band of small conical teeth on each palatine.

Dorsal fin not divided to base at junction of spinous and soft-rayed portions, but fin may appear notched at junction. Distal margin of anal fin rounded. Pectoral fin symmetrical, middle rays longest. Caudal fin forked; principal rays 15 (8 + 7); branched rays 13 (7 + 6). Vertebrae 26 (11 + 15; see **Remarks**). Parapophyses present on first caudal vertebra. Formula for configuration of supraneural bones, etc. 0/0/2/1+1/1/ (Fig. 2B). Pleural ribs on vertebrae 3 through 11. Epineurals associated with first 19 vertebrae. No trisegmental pterygiophores associated with dorsal and anal fins.

Lateral line complete, extending to base of caudal fin (running in almost a straight line parallel to dorsal body contour a few scale rows below dorsal fin, curving gently to near midlateral axis of body on caudal peduncle). Tubes in lateral-line scales simple. No secondary squamation. Maxilla, interorbital region, lachrymal, cheek, preopercle, interopercle, opercle, and subopercle densely covered with scales; scales on dorsum of snout not reaching anterior end of snout—leaving considerable area scaleless; most of lateral aspect of snout naked; no scales on lower jaw, gular region, branchiostegals, and branchiostegal membranes. Dorsal and anal fins without scales, but each with low scaly sheath at its base. Squamation well developed on bases of pectoral, pelvic, and caudal fins and continuing onto fins. Well-developed axillary process of modified scales at pelvic-fin base. Modified scales (interpelvic process) overlapping pelvic-fin bases along midventral line. Other characters are those of the single species.

Remarks. The genus *Anatolanthias* is monotypic. In the original description of the genus, Anderson et al. (1990) gave the vertebral count as 26 (10 + 16). Examination of a new radiograph of the holotype of *Anatolanthias apiomycter* clearly shows the count to be 11 + 15, rather than 10 + 16. Anderson et al. (1990) hypothesized that the genera *Luzonichthys*, *Rabaulichthys*, and *Anatolanthias* constitute a monophyletic assemblage based on their sharing three apparently derived traits—anterior naris rather remote from posterior naris, vomerine dentition absent or extremely reduced, and numbers of pairs of epineurals 16 to 19. Another character state that may be a synapomorphy for those genera is the possession of nine pairs of pleural ribs.

Anatolanthias apiomycter Anderson, Parin, and Randall, 1990

Nazca Gemfish

Fig. 32; Tables 2–6; Map 3

Anatolanthias apiomycter Anderson, Parin, and Randall, 1990:924, figs.1, 2 (original description, illustrations; holotype USNM 309202, 93.9 mm SL; type locality eastern South Pacific, ca. 1500 kilometers off coast of Chile, near southwest end of Nazca Ridge).—Rojas and Pequeño, 1998a:176 (species account after Anderson et al., 1990).

Diagnosis. As for the genus.

Description. Fleshy papillae on posterior half of orbital border 20 to 22. Dorsal-fin rays X, 16. Anal-fin rays III, 7. Pectoral-fin rays 21 or 22. Gillrakers 10 or 11 + 26 or 27—total 37. Lateral-line scales 62 or 63. Procurrent caudal-fin rays 14, dorsally and ventrally.

Head length 27 to 28% SL. Snout length 5 to 6% SL. Orbit diameter 8 to 9% SL. Body depth at first dorsal-fin spine 24 to 25% SL. Fins without elongated spines or soft rays. Longest dorsal spine (fifth or sixth) ca. 12% SL. Depressed anal-fin length 28 to 29% SL. Pelvic-fin length 22 to 24% SL. Caudal fin forked (lobes intact only on holotype). Upper caudal-fin lobe ca. 34 SL. Lower caudal-fin lobe ca. 32% SL.

Coloration. Anderson et al. (1990:926) wrote: "A color transparency of the holotype, taken shortly after capture, shows: body uniformly red, iris red, dorsal and caudal fins red, anal fin paler, paired fins not clearly visible."

Distribution. *Anatolanthias apiomycter* is known only from the type locality (25°41.7'S, 85°23.7'W; 160–168 m) in the eastern South Pacific, near the southwest end of the Nazca Ridge, about 1500 kilometers off the coast of Chile. Parin (1991:679) listed the species in a table of fishes recorded from the Nazca and Sala y Gómez ridges, and Parin et al. (1997:172) noted its occurrence at the type locality, Seamount 12 (Bolshaya), a guyot in the transitional Nazca/Sala y Gómez area. Pequeño (1997:83) included *A. apiomycter* in a list of Chilean fishes.

Material Examined. Known only from the holotype, 93.9 mm SL, and the single paratype, 89.0 mm SL, (see Anderson et al., 1990:924).

LITERATURE CITED

Afonso, P., F. M. Porteiro, R. S. Santos, J. P. Barreiros, J. Worms, and P. Wirtz. 1999. Coastal marine fishes of São Tomé Island (Gulf of Guinea). Bulletin of the University of the Azores, Arquipélago - Life and Marine Sciences 17 A:65–92.

Ahlstrom, E. H., J. L. Butler, and B. Y. Sumida. 1976. Pelagic stromateoid fishes (Pisces, Perciformes) of the eastern Pacific: Kinds, distributions, and early life histories and observations on five of these from the northwest Atlantic. Bulletin of Marine Science 26(3):285–402.

Akhilesh, K. V., N. G. K. Pillai, U. Ganga, K. K. Bineesh, C. P. Rajool Shanis, and H. Manjebrayakath. 2009. First record of the anthiine fish, *Meganthias filiferus* (Perciformes: Serranidae) from Indian waters. Marine Biodiversity Records 2, e113:1–2. Published online.

Allen, G. R. 1976. Descriptions of three new fishes from Western Australia. The Journal of the Royal Society of Western Australia 59 (Part 1):24–30.

Allen, G. R., and J. T. Moyer. 1980. *Ellerkeldia wilsoni*, a new species of serranid fish from southwestern Australia. Japanese Journal of Ichthyology 26(4):329–333.

Allen, G. R., and J. E. Randall. 1990. *Hypoplectrodes cardinalis*, a new name for the serranid fish *H. ruber* (Allen) from southwestern Australia, with discussion of *H. semicinctus* from Juan Fernandez Islands. Revue Française d'Aquariologie 17:45–46.

Allen, G. R., and D. R. Robertson. 1994. Fishes of the tropical eastern Pacific. University of Hawaii Press, Honolulu, xx + 332 pp., many color illustrations.

Allué, C., D. Lloris, and S. Meseguer. 2000. Colecciones biológicas de referencia (1982–1999) del Instituto de Ciencias del Mar (CSIC): Catálogo de peces. Consejo Superior de Investigaciones Cientificas, Barcelona, pp. 1–198.

Anderson, M. E., and R. H. Rosenblatt. 1989. *Pacificogramma stepanenkoi* Kharin, 1983 (family Grammatidae), a junior synonym of *Pronotogrammus multifasciatus* Gill, 1863 (family Serranidae). California Fish and Game 75(2):124–125.

Anderson, W. D., Jr. 2003. Anthiinae. Pp. 1316–1318, 1330–1333, 1350–1352, 1363–1364, *in*: P. C. Heemstra, W. D. Anderson, Jr., and P. S. Lobel. Serranidae. Pp. 1308–1369, *in*: K. E. Carpenter (editor). The living marine resources of the western central Atlantic. Vol. 2: Bony fishes part 1 (Acipenseridae to Grammatidae). FAO Species Identification Guide for Fishery Purposes and American Society of Ichthyologists and Herpetologists Special Publication No.5. Food and Agriculture Organization of the United Nations, Rome. [Dated 2002, but published in 2003.]

Anderson, W. D., Jr. 2006. *Meganthias carpenteri*, new species of fish from the eastern Atlantic Ocean, with a key to eastern Atlantic Anthiinae (Perciformes: Serranidae). Proceedings of the Biological Society of Washington 119(3):404–417.

Anderson, W. D., Jr. 2008. A new species of the perciform fish genus *Plectranthias* (Serranidae: Anthiinae) from the Nazca Ridge in the eastern South Pacific. Proceedings of the Biological Society of Washington 121(4):429–437.

Anderson, W. D., Jr. In press. Anthiinae, *in*: P. C. Heemstra and W. D. Anderson, Jr. Serranidae, *in*: K. E. Carpenter (editor). The living marine resources of the eastern central Atlantic. Bony fishes. FAO Species Identification Guide for Fishery Purposes. Food and Agriculture Organization of the United Nations, Rome.

Anderson, W. D., Jr., and C. C. Baldwin. 2000. A new species of *Anthias* (Teleostei: Serranidae: Anthiinae) from the Galápagos Islands, with keys to *Anthias* and eastern Pacific Anthiinae. Proceedings of the Biological Society of Washington 113(2):369–385.

Anderson, W. D., Jr., and C. C. Baldwin. 2002. *Plectranthias lamillai* Rojas and Pequeño, 1998: A junior synonym of *Plectranthias exsul* Heemstra and Anderson, 1983. Copeia 2002(1):233–238.

Anderson, W. D., Jr., and G. García-Moliner. 2012. A new species of *Odontanthias* Bleeker (Perciformes: Serranidae: Anthiinae) from Mona Passage off Puerto Rico, the first record of the genus from the Atlantic Ocean. aqua, International Journal of Ichthyology 18(1):25–30.

Anderson, W. D., Jr., and P. C. Heemstra. 1980. Two new species of western Atlantic *Anthias* (Pisces: Serranidae), redescription of *A. asperilinguis* and review of *Holanthias martinicensis*. Copeia 1980(1):72–87.

Anderson, W. D., Jr., and P. C. Heemstra. 1989. *Ellerkeldia*, a junior synonym of *Hypoplectrodes*, with redescriptions of the type species of the genera (Pisces: Serranidae: Anthiinae). Proceedings of the Biological Society of Washington 102(4):1001–1017.

Anderson, W. D., Jr., N. V. Parin, and J. E. Randall. 1990. A new genus and species of anthiine fish (Pisces: Serranidae) from the eastern South Pacific with comments on anthiine relationships. Proceedings of the Biological Society of Washington 103(4):922–930.

Anderson, W. D., Jr., and J. E. Randall. 1991. A new species of the anthiine genus *Plectranthias* (Pisces: Serranidae) from the Sala y Gómez Ridge in the eastern South Pacific, with comments on *P. exsul*. Proceedings of the Biological Society of Washington 104(2):335–343.

Anderson, W. D., Jr., and V. G. Springer. 2005. Review of the perciform fish genus *Symphysanodon* Bleeker (Symphysanodontidae), with descriptions of three new species, *S. mona*, *S. parini*, and *S. rhax*. Zootaxa 996:1–44.

Andrew, T. G., T. Hecht, P. C. Heemstra, and J. R. E. Lutjeharms. 1995. Fishes of the Tristan da Cunha Group and Gough Island, South Atlantic Ocean. J. L. B. Smith Institute of Ichthyology, Ichthyological Bulletin No. 63:1–43, 2 plates.

Baldwin, C. C. 1990. Morphology of the larvae of American Anthiinae (Teleostei: Serranidae), with comments on relationships within the subfamily. Copeia 1990(4):913–955.

Baldwin, C. C., and G. D. Johnson. 1993. Phylogeny of the Epinephelinae (Teleostei: Serranidae). Bulletin of Marine Science 52(1):240–283.

Baldwin, C. C., and F. J. Neira. 1998. Serranidae (Anthiinae): Sea basses, sea perches, wirrahs. Pp. 288–293, *in*: F. J. Neira, A. G. Miskiewicz, and T. Trnski (editors), Larvae of temperate Australian fishes: Laboratory guide for larval fish identification. University of Western Australia Press, Nedlands, Western Australia, pp. i–xx + 1–474.

Barans, C. A., E. J. Gutherz, and R. S. Jones. 1986. Submersible avoidance by yellowfin bass, *Anthias nicholsi*. Northeast Gulf Science 8(1):91–95.

Barton, O. 1947. Two new fishes, an *Eques* and a *Holanthias*, from Peru. American Museum Novitates No. 1350:1–3, figs. 1, 2.

Bauchot, M.-L. 1987. Poissons osseux. Pp. 891–1422, *in*: W. Fischer, M.-L. Bauchot, and M. Schneider (editors), Fiches FAO d'identification des espèces pour les besoins de la pêche. (Révision 1.) Méditerranée et mer Noire. Zone de pêche 37. Vol. 2. Vertébrés. FAO, Rome, pp. 761–1530.

Bauchot, M. L., M. Desoutter, and J. E. Randall. 1984. Catalogue critique des types de poissons du Muséum national d'Histoire naturelle (Famille des Serranidae). Bulletin du Muséum national d'Histoire naturelle, Paris, 4th Series, Vol. 6, Section A, No. 3, Supplement:3–82.

Bean, T. H. 1912. Description of new fishes of Bermuda. Proceedings of the Biological Society of Washington 25: 121–126.

Béarez, P., and P. Jiménez Prado. 2003. New records of serranids (Perciformes) from the continental shelf of Ecuador with a key to the species, and comments on ENSO-associated fish dispersal. Cybium 27(2):107–115.

Bellotti, C. 1879. Note ittiologiche. Osservazioni fatte sulla collezione ittiologica del Civico Museo di Storia Naturale in Milano. Atti della Società Italiana di Scienza Naturali 22: 33–38.

Beltrán-León, B. S., and R. Ríos Herrera. 2000. Estadios tempranos de peces del Pacifico Colombiano. Instituto Nacional de Pesca y Acuicultura-INPA, Buenaventura, Colombia. Two volumes, consecutively paginated: 1–359 and 360–727.

Berrisford, C. D. 1969. Biology and zoogeography of the Vema Seamount: A report on the first biological collection made on the summit. Transactions of the Royal Society of South Africa 38 (Part 4):387–398.

Bianchi, C. N., R. Haroun, C. Morri, and P. Wirtz. 2000. The subtidal epibenthic communities off Puerto del Carmen (Lanzarote, Canary Islands). Bulletin of the University of the Azores, Arquipélago - Life and Marine Sciences, Supplement 2(Part A):145–155.

Bleeker, P. 1873. Sur les espèces indo-archipélagique d'*Odontanthias* et de *Pseudopriacanthus*. Nederlandsch Tijdschrift voor de Dierkunde 4:235–240.

Bleeker, P. 1874. Sur les espèces insulindiennes de la famille des Cirrhitéoïdes. Verhandelingen der Koninklijke Akademie van Wetenschappen (Amsterdam) 15:1–20.

Bloch, M. E. 1792. Naturgeschichte der ausländischen Fische. Vol. 6. Berlin, pp. i–xii + 1–126, plates 289–323.

Boulenger, G. A. 1895. Catalogue of the perciform fishes in the British Museum. 2nd ed. London, pp. i–xx + 1–394, plates 1–15.

Briggs, J. C., H. D. Hoese, W. F. Hadley, and R. S. Jones. 1964. Twenty-two new marine fish records for the northwestern Gulf of Mexico. The Texas Journal of Science 16(1):113–116.

Brito, A. 1991. Catálogo de los peces de las Islas Canarias. Francisco Lemus, La Laguna, Tenerife, Canary Islands, 230 pp.

Brito, A., P. J. Pascual, J. M. Falcón, A. Sancho, and G. González. 2002. Peces de las Islas Canarias: Catálogo comentado e ilustrado. Francisco Lemus, Arafo, Tenerife, Canary Islands, 419 pp., many color photos.

Bubley, W. J., and O. Pashuk. 2010. Life history of a simultaneously hermaphroditic fish, *Diplectrum formosum*. Journal of Fish Biology 77:676–691.

Bullock, L. H., and M. F. Godcharles. 1982. Range extensions for four sea basses (Pisces: Serranidae) from the eastern Gulf of Mexico with a color note on *Hemanthias leptus* (Ginsburg). Northeast Gulf Science 5(2):53–56.

Bullock, L. H., and G. B. Smith. 1991. Seabasses (Pisces: Serranidae). Memoirs of the Hourglass Cruises, 8(Part 2): 1–243, plates 1–19.

Burgess, G. H., G. W. Link, Jr., and S. W. Ross. 1979. Additional marine fishes new or rare to Carolina waters. Northeast Gulf Science 3(2):74–87.

Burgess, W., and H. R. Axelrod. 1973. Pacific marine fishes. Book 2. T. F. H. Publications, Neptune City, New Jersey, pp. 281–560, plates 1–529.

Bussing, W. A., and M. I. López. 1994. Demersal and pelagic inshore fishes of the Pacific coast of lower Central America: An illustrated guide. Revista de Biología Tropical, Special Publication, pp. 1–164. [In Spanish and English.]

Bussing, W. A., and M. I. López. 2005. Fishes of Cocos Island and reef fishes of the Pacific coast of lower Central America. Revista de Biología Tropical 53 (Suppl. 2):1–192. [In Spanish and English.]

Carvalho-Filho, A. 1999. Peixes: Costa brasileira. 3rd edition. Melro, São Paulo.

Carvalho-Filho, A., C. E. L. Ferreira, and M. Craig. 2009. A shallow water population of *Pronotogrammus martinicensis* (Guichenot, 1868) (Teleostei: Serranidae: Anthiinae) from south-western Atlantic, Brazil. Zootaxa 2228:29–42.

Castelnau, F. L. de. 1861. Mémoire sur les poissons de l'Afrique australe. Paris, pp. i–vii + 1–78.

Castelnau, F. de. 1879. Essay on the ichthyology of Port Jackson. The Proceedings of the Linnean Society of New South Wales 3 (Pt. 4):347–402.

Cervigón, F. 1966. Los peces marinos de Venezuela. Tomo 1. Fundación La Salle de Ciencias Naturales, Caracas, pp. 1–436.

Cervigón, F. 1991. Los peces marinos de Venezuela. 2nd ed. Vol. 1. Fundación Científica Los Roques, Caracas, pp. 1–425, 6 color plates.

Chakrabarty, P. 2010. The transitioning state of systematic ichthyology. Copeia 2010(3):513–515.

Chirichigno F., N. 1974. Clave para identificar los peces marinos del Peru. Instituto del Mar del Peru, Informe No. 44:1–387, Addenda: 2 pp.

Claro, R., R. G. Gilmore, C. R. Robins, and J. E. McCosker. 2000. Nuevos registros para la ictiofauna marina de Cuba. Avicennia: Revista de Ecología, Oceanología y Biodiversidad Tropical 12/13:19–24.

Coleman, F. 1981. Protogynous hermaphroditism in the anthiine serranid fish *Holanthias martinicensis*. Copeia 1981(4):893–895.

Coleman, F. 1982. Redescription, osteology, sexuality, and variation in the western Atlantic anthiine fish *Holanthias martinicensis* (Serranidae). Master's Thesis, College of Charleston, Charleston, South Carolina, pp. i–viii + 1–92.

Coleman, F. 1983. *Hemanthias peruanus*, another hermaphroditic anthiine serranid. Copeia 1983(1):252–253.

Colin, P. L. 1974. Observation and collection of deep-reef fishes off the coasts of Jamaica and British Honduras (Belize). Marine Biology 24:29–38.

Collette, B. B. 1962. *Hemiramphus bermudensis*, a new halfbeak from Bermuda, with a survey of endemism in Bermudian shore fishes. Bulletin of Marine Science of the Gulf and Caribbean 12(3):432–449.

Collette, B. B., and N. V. Parin. 1991. Shallow-water fishes of Walters Shoals, Madagascar Ridge. Bulletin of Marine Science 48(1):1–22.

Cossman, M. 1889. Catalogue illustré des coquilles fossiles de l'Éocène des environs de Paris. Annales de la Société Royale Malacologique de Belgique 24:3–381.

Craig, M. T., and P. A. Hastings. 2007. A molecular phylogeny of the groupers of the subfamily Epinephelinae (Serranidae) with a revised classification of the Epinephelini. Ichthyological Research 54:1–17.

Cuvier, G. 1828. *In*: G. Cuvier and A. Valenciennes, Histoire naturelle des poissons. Tome 2. Paris, pp. i–xxiv + 1–490, plates 9–40.

De Buen, F. 1959. Lampreas, tiburones, rayas y peces en la Estacion de Biologia Marina de Montemar, Chile. Revista de Biologia Marina 9 (1–3):3–200.

Dennis, G. D., and T. J. Bright. 1988a. Reef fish assemblages on hard banks in the northwestern Gulf of Mexico. Bulletin of Marine Science 43(2):280–307.

Dennis, G. D., and T. J. Bright. 1988b. New records of fishes in the northwestern Gulf of Mexico, with notes on some rare species. Northeast Gulf Science 10(1):1–18.

Dooley, J. K., J. van Tassell, and A. Brito. 1985. An annotated checklist of the shore-fishes of the Canary Islands. American Museum Novitates 2824:1–49, figs 1–5.

Duhamel, G. 1984. Ichtyofaune d'un haut-fond (34°54'S, 53°14'E) de l'océan Indien sud-ouest. Cybium 8(4):91–94.

Duhamel, G. 1989. Ichtyofaune des îles Saint-Paul et Amsterdam (océan Indien Sud). Mésogée, Bulletin du Museum d'Histoire Naturelle de Marseille, 49:21–47. [Dated March 1990 at end of document.]

Duhamel, G. 1997. L'ichtyofaune des îles australes françaises de l'océan Indien. Cybium 21(1, suppl.):147–168.

Edwards, A. 1990. Fish and fisheries of Saint Helena Island. Centre for Tropical Coastal Management Studies. University of Newcastle upon Tyne, England, pp. i–viii + 1–152, 24 plates, 94 figs.

Edwards, A. J. 1993. New records of fishes from the Bonaparte Seamount and Saint Helena Island, South Atlantic. Journal of Natural History 27:493–503.

Edwards, A. J., A. C. Gill, and P. O. Abohweyere. 2003. A revision of F. R. Irvine's Ghanaian marine fishes in the collections of The Natural History Museum, London. Journal of Natural History 37(18):2213–2267.

Edwards, A. J., and C. W. Glass. 1987. The fishes of Saint Helena Island, South Atlantic Ocean. I. The shore fishes. Journal of Natural History 21:617–686.

Edwards, A., and R. Lubbock. 1983. The ecology of Saint Paul's Rocks (equatorial Atlantic). Journal of Zoology, Proceedings of the Zoological Society of London, 200, Pt. 1:51–69.

Erisman, B. E., M. T. Craig, and P. A. Hastings. 2009. A phylogenetic test of the size-advantage model: Evolutionary changes in mating behavior influence the loss of sex change in a fish lineage. The American Naturalist 174(3):E83–E99. [E-Article.]

Erisman, B. E., and P. A. Hastings. 2011. Evolutionary transitions in the sexual patterns of fishes: Insights from a phylogenetic analysis of the seabasses (Teleostei: Serranidae). Copeia 2011(3):357–364.

Erisman, B. E., J. A. Rosales-Casián, and P. A. Hastings. 2008. Evidence of gonochorism in a grouper, Mycteroperca rosacea, from the Gulf of California, Mexico. Environmental Biology of Fishes 82:23–33. [Published online 02 May 2007.]

Eschmeyer, W. N. 1990. Catalog of the genera of recent fishes. California Academy of Sciences, San Francisco, pp. i–vi + 1–697.

Eschmeyer, W. N. (editor). 1998. Catalog of fishes. 3 Vols. Special Publication No. 1, Center for Biodiversity Research and Information, California Academy of Sciences, San Francisco, pp. 1–2905.

Eschmeyer, W. N. (editor). Catalog of Fishes electronic version (updated 12 January 2012). http://research.calacademy.org/ichthyology/catalog/fishcatmain.asp.

Feitoza, B. M., L. A. Rocha, O. J. Luiz-Júnior, S. R. Floeter, and J. L. Gasparini. 2003. Reef fishes of St. Paul's Rocks: New records and notes on biology and zoogeography. aqua, Journal of Ichthyology and Aquatic Biology 7(2):61–82.

Firth, F. E. 1933. *Anthias nicholsi*, a new fish taken off Virginia in the deep-water trawl fishery. Copeia 1933(4):158–160.

Fishelson, L. 1970. Protogynous sex reversal in the fish *Anthias squamipinnis* (Teleostei, Anthiidae) regulated by the presence or absence of a male fish. Nature 227:90–91.

Fitch, J. E. 1982. Revision of the eastern North Pacific anthiin basses (Pisces: Serranidae). Natural History Museum of Los Angeles County, Contributions in Science No. 339:1–8.

Fowler, H. W. 1925. Fishes from Natal, Zululand, and Portuguese East Africa. Proceedings of the Academy of Natural Sciences of Philadelphia 77:187–268.

Fowler, H. W. 1937. Notes on fishes from the Gulf Stream and the New Jersey coast. Proceedings of the Academy of Natural Sciences of Philadelphia 89:297–308.

Francis, M. 1988. Coastal fishes of New Zealand: A diver's identification guide. Heinemann Reed, Auckland, New Zealand, pp. 1–63, plates 1–146.

Garman, S. 1899. The fishes. *In*: Reports on an exploration off the west coasts of Mexico, Central and South America, and off the Galapagos Islands, in charge of Alexander Agassiz, by the U.S. Fish Commission Steamer "Albatross," during 1891, Lieut. Commander Z. L. Tanner, U. S. N., commanding. No. XXVI. Memoirs of the Museum of Comparative Zoölogy at Harvard College 24: Text: pp. 1–431, Atlas: plates 1–85 + A-M.

Gilbert, C. H. 1890. A preliminary report on the fishes collected by the steamer Albatross on the Pacific coast of North America during the year 1889, with descriptions of twelve new genera and ninety-two new species. Proceedings of the United States National Museum 13(No. 797):49–126.

Gilhen, J., and D. E. McAllister. 1981. First Canadian record of yellowfin bass, *Anthias nicholsi* Firth, taken off Nova Scotia. Proceedings of the Nova Scotian Institute of Science 31:251–254.

Gill, T. 1862. Remarks on the relations of the genera and other groups of Cuban fishes. Proceedings of the Academy of Natural Sciences of Philadelphia 14:235–242.

Gill, T. 1863. Catalogue of the fishes of Lower California, in the Smithsonian Institution, collected by Mr. J. Xantus. Part IV. Proceedings of the Academy of Natural Sciences of Philadelphia 15(2):80–88.

Gines, H., and F. Cervigón. 1968. Exploracion pesquera en las costas de Guayana y Surinam año 1967. Memoria de la Sociedad de Ciencias Naturales La Salle 28(79):5–96.

Ginsburg, I. 1952. Eight new fishes from the Gulf Coast of the United States, with two new genera and notes on geographic distribution. Journal of the Washington Academy of Sciences 42(3):84–101.

Ginsburg, I. 1954. Four new fishes and one little-known species from the east coast of the United States including the Gulf of Mexico. Journal of the Washington Academy of Sciences 44(8):256–264.

Gosline, W. A. 1966. The limits of the fish family Serranidae, with notes on other lower percoids. Proceedings of the California Academy of Sciences, 4th Series, 33(6):91–111.

Grove, J. S., and R. J. Lavenberg. 1997. The fishes of the Galápagos Islands. Stanford Univ. Press, Stanford, California, pp. i–xlvi + 1–863, color photos. 1–151.

Guichenot, A. 1868. Index generum ac specierum anthiadidorum hucusque in Museo Parisiensis observatorum. Annales de la Société Linnéenne du Département de Maine-et-Loire 10:80–87.

Günther, A. 1859. Catalogue of the acanthopterygian fishes in the collection of the British Museum. Vol. 1. London, pp. i–xxxii + 1–524.

Günther, A. 1868. Report on a collection of fishes made at St. Helena by J. C. Melliss, Esq. Proceedings of the Zoological Society of London 1868:225–228, plates 18 & 19.

Günther, A. 1872. Report on several collections of fishes recently obtained for the British Museum. Proceedings of the Zoological Society of London 1871 (Pt. 3):652–675, plates 53–70.

Gutherz, E. J., W. R. Nelson, R. S. Jones, C. A. Barans, C. A. Wenner, G. M. Russell, and A. K. Shah. 1995. Population estimates of deep-water finfish species based on submersible observations and intensive fishing efforts off Charleston, S. C. Pp. 103–123, in: J. A. Bohnsack and A. Woodhead (compilers). Proceedings of the 1987 SEAMAP Passive Gear Assessment Workshop at Mayaguez, Puerto Rico. NOAA Technical Memorandum NMFS-SEFSC-365, pp. i–vi + 1–177.

Hastings, P. A. 1981. Gonad morphology and sex succession in the protogynous hermaphrodite *Hemanthias vivanus* (Jordan and Swain). Journal of Fish Biology 18(4):443–454.

Hector, J. 1875. Notes on New Zealand ichthyology. Transactions and Proceedings of the New Zealand Institute 7 (for 1874) (Art. 34):239–250, plates 10 & 11.

Heemstra, P. C. 1973. *Anthias conspicuus* sp. nova (Perciformes: Serranidae) from the Indian Ocean, with comments on related species. Copeia 1973(2): 200–210.

Heemstra, P. C. 1995. Serranidae. Pp. 1565–1613, in: W. Fischer, F. Krupp, W. Schneider, C. Sommer, K. E. Carpenter, and V. H. Niem (editors). Guía FAO para la identificación de especies para los fines de la pesca. Pacifico centro-oriental. Vol. III. Vertebrados - Parte 2:1201–1813. Food and Agriculture Organization of the United Nations, Rome.

Heemstra, P. C. 2010. Taxonomic review of the perciform fish genus *Acanthistius* from the east coast of southern Africa, with description of a new species and designation of a neotype for *Serranus sebastoides* Castelnau, 1861. Zootaxa 2352:59–68.

Heemstra, P. C., and W. D. Anderson, Jr. 1983. A new species of the serranid fish genus *Plectranthias* (Pisces: Perciformes) from the southeastern Pacific Ocean, with comments on the genus *Ellerkeldia*. Proceedings of the Biological Society of Washington 96(4):632–637.

Heemstra, P. C., and J. E. Randall. 1986. Family No. 166: Serranidae. Pp. 509–537, plates 32–45, *in*: M. M. Smith and P. C. Heemstra (editors), Smith's Sea Fishes. Macmillan South Africa, Johannesburg, i–xx + 1–1047.

Heemstra, P. C., and J. E. Randall. 2009. A review of the anthiine fish genus *Plectranthias* (Perciformes: Serranidae) of the western Indian Ocean, with description of a new species, and a key to the species. Smithiana, Bulletin No. 10:3–17, color plate 1.

Herman, J. S., R. Y. McCowan, and G. N. Swinney. 1990. Catalogue of the type specimens of Recent vertebrates in the National Museums of Scotland. National Museums of Scotland Information Series No. 4:i–vi + 1–34.

Hildebrand, S. F., and O. Barton. 1949. A collection of fishes from Talara, Perú. Smithsonian Miscellaneous Collections 111(10):1–36.

Hobson, E. S. 1975. First California record of the serranid fish *Anthias gordensis* Wade. California Fish and Game 61(2):111–112.

Hoogesteger, J. 1988. The potential for offshore fisheries in the St. Helena exclusive fishing zone. Overseas Development Administration, London, 93 pp. + 8 appendices.

Houde, E. D. 1982. Kinds, distributions and abundances of sea bass larvae (Pisces: Serranidae) from the eastern Gulf of Mexico. Bulletin of Marine Science 32(2):511–522.

Hovey, T. E., and L. G. Allen. 2000. Reproductive patterns of six populations of the Spotted Sand Bass, *Paralabrax maculatofasciatus*, from southern and Baja California. Copeia 2000(2):459–468.

Hughes, D. R. 1981. Development and organization of the posterior field of ctenoid scales in the Platycephalidae. Copeia 1981(3):596–606.

ICZN: International Commission on Zoological Nomenclature. 1985. International code of zoological nomenclature. 3rd ed. University of California Press, Berkeley, pp. i–xx + 1–338.

ICZN: International Commission on Zoological Nomenclature. 1999. International code of zoological nomenclature. 4th ed. International Trust for Zoological Nomenclature, London, pp. i–xxx + 1–306.

Imamura, H., and M. Yabe. 2002. Demise of the Scorpaeniformes (Actinopterygii: Percomorpha): An alternative phylogenetic hypothesis. Bulletin of Fisheries Sciences Hokkaido University 53(3):107–128.

Irigoyen, A. J., L. C. Gerhardinger, and A. Carvalho-Filho. 2008. On the status of the species of *Acanthistius* (Gill, 1862) (Percoidei) in the south-west Atlantic Ocean. Zootaxa 1813:51–59.

IUCN 2011. IUCN Red List of Threatened Species. Version 2011.2. www. iucnredlist .org. Downloaded on 23 January 2012.

Jenyns, L. 1840–1842. Fish. *In*: The zoology of the voyage of H. M. S. Beagle, under the command of Captain Fitzroy, R. N., during the years 1832 to 1836. Smith, Elder, and Co., London. Issued in 4 parts, pp. i–xvi + 1–172, plates 1–29.

Johnson, G. D. 1975. The procurrent spur: An undescribed perciform caudal character and its phylogenetic implications. Occasional Papers of the California Academy of Sciences 121:1–23.

Johnson, G. D. 1983. *Niphon spinosus*: A primitive epinepheline serranid, with comments on the monophyly and intrarelationships of the Serranidae. Copeia 1983(3):777–787.

Johnson, G. D. 1984. Percoidei: Development and relationships. Pp. 464–498, *in*: H. G. Moser et al. (editors), Ontogeny and systematics of fishes. American Society of Ichthyologists and Herpetologists, Special Publication No. 1. Allen Press, Inc., Lawrence, Kansas, pp. i–x + 1–760.

Johnson, G. D. 1988. *Niphon spinosus*, a primitive epinepheline serranid: Corroborative evidence from the larvae. Japanese Journal of Ichthyology 35(1):7–18.

Johnson, G. D., and C. Patterson. 1993. Percomorph phylogeny: A survey of acanthomorphs and a new proposal. Bulletin of Marine Science 52(1):554–626.

Johnson, J. Y. 1890. On some new species of fishes from Madeira. Proceedings of the Zoological Society of London 1890:452–459.

Jones, G. P. 1980. Contribution to the biology of the redbanded perch, *Ellerkeldia huntii* (Hector), with a discussion on hermaphroditism. Journal of Fish Biology 17:197–207.

Jones, L. L. C., R. R. Johnson, J. Hopkins, and S. M. White. 1985. Additional records of *Pronotogrammus multifasciatus* and *Gempylus serpens* from California. California Fish and Game 71(2):116–117.

Jordan, D. S. 1890. *In*: D. S. Jordan and C. H. Eigenmann, a review of the genera and species of Serranidae found in the waters of America and Europe. Bulletin of the United States Fish Commission 8(for 1888):329–441, plates 60–69.

Jordan, D. S. 1907. *In*: D. S. Jordan and A. W. Herre, a review of the cirrhitoid fishes of Japan. Proceedings of the United States National Museum 33(No. 1562):157–167.

Jordan, D. S., and C. H. Eigenmann. 1890. A review of the genera and species of Serranidae found in the waters of America and Europe. Bulletin of the United States Fish Commission 8(for 1888):329–441, plates 60–69.

Jordan, D. S., and B. W. Evermann. 1896. The fishes of North and Middle America: A descriptive catalogue of the species of fish-like vertebrates found in the waters of North America, north of the Isthmus of Panama. Part I. Bulletin of the United States National Museum No. 47:i–lx + 1–1240.

Jordan, D. S., and B. W. Evermann. 1903. Descriptions of new genera and species of fishes from the Hawaiian Islands. Bulletin of the United States Fish Commission 22(for 1902):161–208.

Jordan, D. S., and R. E. Richardson. 1910. A review of the Serranidae or seabass of Japan. Proceedings of the United States National Museum 37(No. 1714):421–474.

Jordan, D. S., and A. Seale. 1906. Descriptions of six new species of fishes from Japan. Proceedings of the United States National Museum 30(No. 1445):143–148.

Jordan, D. S., and J. Swain. 1885. Description of three new species of fishes (*Prionotus stearnsi*, *Prionotus ophryas*, and *Anthias vivanus*) collected at Pensacola, Florida, by Mr. Silas Stearns. Proceedings of the United States National Museum 7:541–545.

Katayama, M. 1960. Fauna Japonica Serranidae (Pisces). Biogeographical Society of Japan, Tokyo, pp. i–viii + 1–189, plates 1–86.

Katayama, M. 1964. A new genus and species of anthinid fish from Sagami Bay, Japan. Bulletin of the Faculty of Education, Yamaguchi University 13(Pt. 2) (1963):27–33.

Katayama, M. 1975. *Caprodon unicolor*, a new anthiine fish from the North Pacific Ocean. Japanese Journal of Ichthyology 22(1):13–15.

Katayama, M., and K. Amaoka. 1986. Two new anthiine fishes from the eastern tropical Atlantic. Japanese Journal of Ichthyology 33(3):213–222.

Katayama, M., and E. Fujii. 1982. Two new species of the anthiine genus *Lepidoperca* from Australia and New Zealand. Japanese Journal of Ichthyology 29(3):241–252.

Kendall, A. W., Jr. 1979. Morphological comparisons of North American sea bass larvae (Pisces: Serranidae). NOAA Technical Report NMFS Circular 428:i–iv + 1–50.

Kendall, A. W., Jr. 1984. Serranidae: Development and relationships. Pp. 499–510, *in*: H. G. Moser et al. (editors), Ontogeny and systematics of fishes. American Society of Ichthyologists and Herpetologists, Special Publication No. 1. Allen Press, Inc., Lawrence, Kansas, pp. i–x + 1–760.

Kharin, V. Ye. 1983a. *In*: V. Ye. Kharin and V. A. Dudarev. A new species of the genus *Caprodon* Temminck et Schlegel, 1843 (Serranidae) and some remarks on the composition of the genus. Journal of Ichthyology 23(1):20–25. [English translation of the original Russian in Voprosy Ikhtiologii.]

Kharin, V. E. 1983b. Novyi rod i vid grammovykh okunei iz vod yuzhnoi Kalifornii (Osteichthyes, Grammidae). [A new genus and species of grammid perch from the waters of Lower California (Osteichthyes, Grammidae).] Izvestiya Tikhookeanskogo Nauchno-Issledovatel'Skogo Instituta Rybnogo Khozyaistva i Okeanografii 107:116–119. [Bulletin of the Pacific Scientific Institute of Fisheries and Oceanography; in Russian.]

Kharin, V. Ye., and V. A. Dudarev. 1983. A new species of the genus *Caprodon* Temminck et Schlegel, 1843 (Serranidae) and some remarks on the composition of the genus. Journal of Ichthyology 23(1):20–25. [English translation of the original Russian in Voprosy Ikhtiologii.]

Kon, T., T. Yoshino, and Y. Sakurai. 2000. A new anthiine fish (Perciformes: Serranidae), *Holanthias kingyo*, from the Ryukyu Islands. Ichthyological Research 47(1):75–79.

Kuiter, R. H. 1993. Coastal fishes of south-eastern Australia. University of Hawaii Press, Honolulu, pp. i–xxxii + 1–437, many color figs.

Kuiter, R. H. 2004. Basslets, hamlets, and their relatives: A comprehensive guide to selected Serranidae and Plesiopidae. The Marine Fish Families Series. TMC Publishing, Chorleywood, United Kingdom, pp. 1–216, many color figs.

Lavenda, N. 1949. Sexual differences and normal protogynous hermaphroditism in the Atlantic Sea Bass, *Centropristes striatus*. Copeia 1949(3):185–194.

Leviton, A. E., R. H. Gibbs, Jr., E. Heal, and C. E. Dawson. 1985. Standards in herpetology and ichthyology: Part I. Standard symbolic codes for institutional resource collections in herpetology and ichthyology. Copeia 1985(3):802–832.

Lindquist, D. G., and I. E. Clavijo. 1994. Quantifying deep reef fishes from a submersible and notes on a live collection and diet of the Red Barbier, *Hemanthias vivanus*. The Journal of the Elisha Mitchell Scientific Society 109(3):135–140. [Dated 1993, but published on 22 March 1994.]

Linnaeus, C. 1758. Systema naturae. Vol. I. Regnum animale. 10th ed., Holmiae, 824 pp. [Photographic facsimile, 1956, printed by order of the Trustees, British Museum (Natural History), London.]

Longley, W. H. 1935. Osteological notes and descriptions of new species of fishes. Carnegie Institution of Washington Year Book No. 34:86–89.

Longley, W. H., and S. F. Hildebrand. 1940. New genera and species of fishes from Tortugas, Florida. Papers from Tortugas Laboratory, Carnegie Institution of Washington 32:223–285, plate 1.

Lowe, R. T. 1843. A history of the fishes of Madeira, with original figures from nature of all the species, by Hon. C. E. C. Norton and M. Young. London, 1843–60. Pp. 1–196, 27 plates. [Pt. 1, July 1843:i–xvi + 1–20, plates I–IV; Pt. 2, Sep. 1843:21–52, plates V–VII; Pt. 3, Nov. 1843:53–84, plates IX–XII; Pt. 4, Jan. 1844:85–116, plates XIII–XVII; Pt. 5, Oct. 1860:117–196.]

Lubbock, R., and A. Edwards. 1981. The fishes of Saint Paul's Rocks. Journal of Fish Biology 18:135–157.

Luiz Jr., O. J., J. C. Joyeux, and J. L. Gasparini. 2007. Rediscovery of *Anthias salmopunctatus* Lubbock & Edwards, 1981, with comments on its natural history and conservation. Journal of Fish Biology 70:1283–1286.

Mabee, P. M. 1988. Supraneural and predorsal bones in fishes: Development and homologies. Copeia 1988(4):827–838.

Markle, D. F., W. B. Scott, and A. C. Kohler. 1980. New and rare records of Canadian fishes and the influence of hydrography on resident and nonresident Scotian Shelf ichthyofauna. Canadian Journal of Fisheries and Aquatic Sciences 37(1):49–65.

Matsuura, K. 1983. Serranidae. Pp. 299–315, *in*: T. Uyeno, K. Matsuura, and E. Fujii (editors). Fishes trawled off Suriname and French Guiana. Japan Marine Fishery Resource Research Center, Tokyo, pp. 1–519, many color figs.

Maugé, L. A. 1990. Anthiidae. Pp. 707–708, *in*: J. C. Quéro, J. C. Hureau, C. Karrer, A. Post, and L. Saldanha (editors), Check-list of the fishes of the eastern tropical Atlantic, *Clofeta*. European Ichthyological Union, Paris, Vol. 2, pp. 520–1080.

Maul, G. E. 1976. The fishes taken in bottom trawls by R. V. "Meteor" during the 1967 seamount cruises in the northeast Atlantic. "Meteor" Forsch.-Ergebnisse, Series D, No. 22:1–69.

McBride, R. S., K. J. Sulak, P. E. Thurman, and A. K. Richardson. 2010. Age, growth, mortality, and reproduction of Roughtongue Bass, *Pronotogrammus martinicensis* (Serranidae), in the northeastern Gulf of Mexico. Gulf of Mexico Science 27(1):30–38. [Dated 2009, but published in 2010.]

McCosker, J. E., G. Merlen, D. J. Long, R. G. Gilmore, and C. Villon. 1997. Deepslope fishes collected during the 1995 eruption of Isla Fernandina, Galápagos. Noticias de Galápagos No. 58: 22–26.

McCosker, J. E., and R. H. Rosenblatt. 2010. The fishes of the Galápagos Archipelago: An update. Proceedings of the California Academy of Sciences, Series 4, Vol. 61, Supplement II, No. 11:167–195.

Meléndez, R., and C. Villalba. 1992. Nuevos registros y antecedentes para la ictiofauna del Archipiélago de Juan Fernández, Chile. Estudios Oceanológicos 11:3–29.

Menezes, G. M., O. Tariche, M. R. Pinho, P. N. Duarte, A. Fernandes, and M. A. Aboim. 2004. Annotated list of fishes caught by the R/V ARQUIPÉLAGO off the Cape Verde Archipelago. Bulletin of the University of the Azores, Arquipélago - Life and Marine Sciences 21A:57–71.

Miranda Ribeiro, A. de. 1903. Pescas do "Annie." Boletim da Sociedade Nacional de Agricultura, Rio de Janeiro, Nos. 4–7:1–53.

Mooi, R. D., and A. C. Gill. 2008. Phylogenies without synapomorphies—a crisis in systematics or what we don't node—the imperative of character evidence for phylogeny reconstruction. *In*: Abstracts of the 88th Meeting of the American Society of Ichthyologists and Herpetologists.

Mooi, R. D., and A. C. Gill. 2010a. Phylogenies without synapomorphies—a crisis in fish systematic: Time to show some character. Zootaxa 2450:26–40.

Mooi, R. D., and A. C. Gill. 2010b. A transitioning state or harmful mutation in systematic ichthyology? A reply to Chakrabarty. Copeia 2010(3):516–519.

Moore, J. A., K. E. Hartel, J. E. Craddock, and J. K. Galbraith. 2003. An annotated list of deepwater fishes from off the New England region, with new area records. Northeastern Naturalist 10(2):159–248.

Moore, S. E., S. A. Hesp, N. G. Hall, and I. C. Potter. 2007. Age and size compositions, growth and reproductive biology of the breaksea cod *Epinephelides armatus*, a gonochoristic serranid. Journal of Fish Biology 71:1407–1429.

Nakamura, I. 1986. *Acanthistius patachonicus* (Jenyns, 1840). Pp. 196–197, *in*: I. Nakamura, T. Inada, M. Takeda, and H. Hatanaka, Important fishes trawled off Patagonia. Japan Marine Fishery Resource Research Center, Tokyo.

Nelson, R. S. 1988. A study of the life history, ecology, and population dynamics of four sympatric reef predators (*Rhomboplites aurorubens*, *Lutjanus campechanus*, Lutjanidae; *Haemulon melanurum*, Haemulidae; and *Pagrus pagrus*, Sparidae) on the East and West Flower Garden Banks, northwestern Gulf of Mexico. Ph. D. Dissertation, North Carolina State University, Raleigh. 197 pp. [Not seen.]

Nichols, J. T. 1920. A contribution to the ichthyology of Bermuda. Proceedings of the Biological Society of Washington 33:59–64.

Parin, N. V. 1991. Fish fauna of the Nazca and Sala y Gomez submarine ridges, the easternmost outpost of the Indo-west Pacific zoogeographic region. Bulletin of Marine Science 49(3):671–683.

Parin, N. V., G. A. Golovan, N. P. Pakhorukov, Yu. I. Sazonov, and Yu. N. Shcherbachev. 1981. Fishes from the Nazca and Sala-y-Gomez underwater ridges collected in cruise of R/V "Ikhtiandr." In: Fishes of the open ocean. Institute of Oceanology, Academy of Sciences of the USSR, Moscow ("1980"):5–18. [In Russian with English abstract on p. 115.]

Parin, N. V., A. N. Mironov, and K. N. Nesis. 1997. Biology of the Nazca and Sala y Gómez Submarine ridges, an outpost of the Indo-west Pacific fauna in the eastern Pacific Ocean: Composition and distribution of the fauna, its communities and history. In: The Biogeography of the Oceans. Advances in Marine Biology 32:145–242.

Parker, R. O., Jr., and S. W. Ross. 1986. Observing reef fishes from submersibles off North Carolina. Northeast Gulf Science 8(1):31–49.

Patterson, C., and G. D. Johnson. 1995. The intermuscular bones and ligaments of teleostean fishes. Smithsonian Contributions to Zoology 559:i–iv + 1–83, plates 1 & 2.

Patzner, R. A., R. S. Santos, P. Ré, and R. D. M. Nash. 1992. Littoral fishes of the Azores: An annotated checklist of fishes observed during the "Expedition Azores 1989." Bulletin of the University of the Azores, Arquipélago - Life and Earth Sciences No. 10:101–111.

Penrith, M. J. 1967. The fishes of Tristan da Cunha, Gough Island, and the Vema Seamount. Annals of the South African Museum 48(Part 22):523–548, plate 21.

Penrith, M. J. 1978. An annotated check-list of the inshore fishes of southern Angola. Cimbebasia, Ser. A, 4(11):179–190.

Pequeño, G. 1997. Peces de Chile. Lista sistemática revisada y comentado: *addendum*. Revista de Biologia Marina y Oceanografía 32(2):77–94.

Pequeño, G., and J. Lamilla. 1996a. Desventuradas Islands, Chile: The easternmost outpost of the Indo-west Pacific zoogeographic region. Revista de Biologia Tropical 44(2):929–931.

Pequeño, G., and J. Lamilla. 1996b. Peces de la familia Serranidae en las Islas Desventuradas, Chile (Osteichthyes, Perciformes). Boletín de la Sociedad de Biología de Concepción 67:23–32.

Pequeño, G., and J. Lamilla. 2000. The littoral fish assemblage of the Desventuradas Islands (Chile) has zoogeographical affinities with the western Pacific. Global Ecology & Biogeography 9:431–437.

Pequeño, G., and S. Sáez. 2000. Los peces litorales del archipiélago de Juan Fernández (Chile): Endemismo y relaciones ictiogeográficas. Investigaciones Marinas, Valparaíso 28:27–37.

Quattrini, A. M., and S. W. Ross. 2006. Fishes associated with North Carolina shelf-edge hardbottoms and initial assessment of a proposed marine protected area. Bulletin of Marine Science 79(1):137–163.

Rafinesque, C. S. 1810. Caratteri di alcuni nuovi generi e nuove specie di animali e piante della Sicilia, con varie osservazioni sopra i medesimi. Pt. 1. Palermo, pp. 5–69.

Randall, J. E. 1980. Revision of the fish genus *Plectranthias* (Serrandidae: Anthiinae) with descriptions of 13 new species. Micronesica 16(1):101–187.

Randall, J. E. 1995. Zoogeographic analysis of the inshore Hawaiian fish fauna. Pp. 193–203, Chapter 11, *in*: J. E. Maragos, M. N. A. Peterson, L. G. Eldredge, J. E. Bardach, and H. F. Takeuchi (editors), Marine and coastal biodiversity in the tropical island Pacific region. Vol. 1. Species systematics and information management priorities. Honolulu, Hawaii.

Randall, J. E. 1996. Two new anthiine fishes of the genus *Plectranthias* (Perciformes: Serranidae), with a key to the species. Micronesica 29(2):113–131.

Randall, J. E. 1998. Zoogeography of shore fishes of the Indo-Pacific region. Zoological Studies 37(4):227–268.

Randall, J. E. 2007. Reef and shore fishes of the Hawaiian Islands. Sea Grant College Program, University of Hawai'i, Honolulu, i–xiv + 1–546, many color figs.

Randall, J. E., and G. R. Allen. 1989. *Pseudanthias sheni*, a new serranid fish from Rowley Shoals and Scott Reef, Western Australia. Revue Française d'Aquariologie 15:73–78.

Randall, J. E., and A. Cea. 2011. Shore fishes of Easter Island. University of Hawaii Press, Honolulu, i–xii + 1–164 pp., many color figs.

Randall, J. E., and A. Cea Egaña. 1984. Native names of Easter Island fishes, with comments on the origin of the Rapanui people. Occasional Papers of Bernice Pauahi Bishop Museum 25(12):1–16.

Randall, J. E., A. Cea E., and R. Meléndez C. 2005. Checklist of shore and epipelagic fishes of Easter Island, with twelve new records. Boletin del Museo Nacional de Historia Natural, Chile 54:41–55.

Randall, J. E., and P. C. Heemstra. 2006. Review of the Indo-Pacific fishes of the genus *Odontanthias* (Serranidae: Anthiinae), with descriptions of two new species and a related new genus. Indo-Pacific Fishes No. 38:1–32, plates 1–8.

Randall, J. E., and P. C. Heemstra. 2008. *Meganthias filiferus*, a new species of anthiine fish (Perciformes: Serranidae), from the Andaman Sea off southwestern Thailand. Phuket Marine Biological Center Research Bulletin 68:5–9. [Dated 2007, but published in 2008.]

Randall, J. E., and D. F. Hoese. 1995. Three new species of Australian fishes of the genus *Plectranthias* (Perciformes: Serranidae: Anthiinae). Records of the Australian Museum 47: 327–335.

Randall, J. E., and J. E. McCosker. 1992. Revision of the fish genus *Luzonichthys* (Perciformes: Serranidae: Anthiinae), with descriptions of two new species. Indo-Pacific Fishes No. 21:1–21, plates 1 & 2.

Regan, C. T. 1908. Report on the marine fishes collected by Mr. J. Stanley Gardiner in the Indian Ocean. The Transactions of the Linnean Society of London, Second Series, Zoology 12 (Pt. 3):217–255, plates 23–32.

Regan, C. T. 1913a. A collection of fishes made by Professor Francisco Fuentes at Easter Island. Proceedings of the Zoological Society of London 1913 (Part 3):368–374, plates 55–60.

Regan, C. T. 1913b. The Antarctic fishes of the Scottish National Antarctic Expedition. Transactions of the Royal Society of Edinburgh 49 (Part 2, No.2):229–292, plates 1–11.

Regan, C. T. 1914. Diagnoses of new marine fishes collected by the British Antarctic ('Terra Nova') expedition. The Annals and Magazine of Natural History (Series 8) 13 (No. 73):11–17.

Reinboth, R. 1963. Natürlicher Geschlechtswechsel bei *Sacura margaritacea* (Hilgendorf) (Serranidae). Annotationes Zoologicae Japonenses 36:173–178.

Reinboth, R. 1964. Inversion du sexe chez *Anthias anthias* (L.) (Serranidae). Vie et Milieu, Supplément 17:499–503.

Rhodes, K. L., and Y. Sadovy. 2002. Reproduction in the camouflage grouper (Pisces: Serranidae) in Pohnpei, Federated States of Micronesia. Bulletin of Marine Science 70(3):851–869.

Richards, W. J., C. C. Baldwin, and A. Röpke. 2006. Serranidae: Sea basses. Pp. 1225–1331, *in*: W. J. Richards (editor). Early stages of Atlantic fishes: An identification guide for the western central North Atlantic. Vol. I. CRC Press, Taylor & Francis Group, Boca Raton, Florida, pp. [i–xx] + 1–1335.

Roa-Varón, A, L. M. Saavedra-Díaz, A. Acero P., L. S. Mejía M., and G. R. Navas S. 2003. Nuevos registros de peces óseos para el Caribe Colombiano de los órdenes Beryciformes, Zeiformes, Perciformes y Tetraodontiformes. Boletín de Investigaciónes Marinas y Costeras 32:3–24.

Roberts, C. D. 1989. A revision of New Zealand and Australian orange perches (Teleostei; Serranidae) previously referred to *Lepidoperca pulchella* (Waite) with description of a new species of *Lepidoperca* from New Zealand. Journal of Natural History 23:557–589.

Roberts, C. D. 1993. Comparative morphology of spined scales and their phylogenetic significance in the Teleostei. Bulletin of Marine Science 52(1):60–113.

Robins, C. R., and W. A. Starck, II. 1961. Materials for a revision of *Serranus* and related fish genera. Proceedings of the Academy of Natural Sciences of Philadelphia 113(11):259–314.

Rojas, J. R., S. Palma, and G. Pequeño. 1998. Food of the grouper *Caprodon longimanus* from Alejandro Selkirk Island, Chile (Perciformes: Serranidae). Revista de Biologia Tropical 46(4):937–942.

Rojas, J. R., and G. Pequeño. 1998a. Revisión taxonómica de los peces de la subfamilia Anthiinae del Pacífico suroriental chileno (Pisces: Serranidae: Anthiinae). Revista de Biología Marina y Oceanografía 33(2):163–198.

Rojas, J. R., and G. Pequeño. 1998b. *Plectranthias lamillai*, a new anthiine fish species (Perciformes, Serranidae) from the Juan Fernández Archipelago, Chile. Scientia Marina 62(3):203–209.

Rojas, J. R., and G. Pequeño. 1998c. Peces serránidos de la Isla Alejandro Selkirk, Archipiélago Juan Fernández, Chile (Pisces Serranidae): Análisis ictiogeográfico. Investigaciones Marinas, Valparaíso 26:41–58.

Ross, S. W., G. W. Link, Jr., and K. A. MacPherson. 1981. New records of marine fishes from the Carolinas, with notes on additional species. Brimleyana No. 6:61–72.

Ross, S. W., and A. M. Quattrini. 2007. The fish fauna associated with deep coral banks off the southeastern United States. Deep-Sea Research, Part I, 54:975–1007.

Sadovy, Y., and P. L. Colin. 1995. Sexual development and sexuality in the Nassau Grouper. Journal of Fish Biology 46:961–976.

Sadovy, Y., and M. L. Domeier. 2005. Perplexing problems of sexual patterns in the fish genus *Paralabrax* (Serranidae, Serraninae). Journal of Zoology, London, 267:121–133.

Sadovy, Y., and D. Y. Shapiro. 1987. Criteria for the diagnosis of hermaphroditism in fishes. Copeia 1987(1):136–156.

Sadovy de Mitcheson, Y., and M. Liu. 2008. Functional hermaphroditism in teleosts. Fish and Fisheries 9:1–43.

Santos, R. S., F. M. Porteiro, and J. P. Barreiros. 1997. Marine fishes of the Azores: An annotated checklist and bibliography: A catalogue of the Azorean marine ichthyodiversity. Bulletin of the University of the Azores, Arquipélago - Life and Marine Sciences, Supplement 1:i–xxviii + 1–244.

Schaldach, W. J., Jr., L. Huidobro Campos, and H. Espinosa Pérez. 1997. Peces marinos. Pp. 463–471, *in*: E. González Soriano, R. Dirzo, and R. C. Vogt (editors), Historia natural de Los Tuxtlas. México, D. F. 647 pp.

Scott, W. B., and M. G. Scott. 1988. Atlantic fishes of Canada. Canadian Bulletin of Fisheries and Aquatic Sciences No. 219: i–xxx + 1–731.

Shapiro, D. Y. 1986. Intra-group home ranges in a female-biased group of sex-changing fish. Animal Behavior 34:865–870.

Shipp, R. L., and T. S. Hopkins. 1978. Physical and biological observations of the northern rim of the De Soto Canyon made from a research submersible. Northeast Gulf Science 2(2):113–121.

Silva, D. 1936. Sobre um serranideo pouco conhecido *Odontanthias asperilingua* Günther. Revista do Departamento Nacional da Producção Animal 3(1–6):187–188.

Smith, C. L. 1971. Secondary gonochorism in the serranid genus *Liopropoma*. Copeia 1971(2):316–319.

Smith, C. L. 1975. The evolution of hermaphroditism in fishes. Pp. 295–310, *in*: R. Reinboth (editor), Intersexuality in the animal kingdom. Springer Verlag, Berlin, Heidelberg, New York.

Smith, C. L. 1981. Anthiidae. 8 pp, *in*: W. Fischer, G. Bianchi, and W. B. Scott (editors), FAO species identification sheets for fishery purposes. Eastern central Atlantic; fishing areas 34, 47 (in part). Vol. 1. Canada Funds-in-Trust. Ottawa, Department of Fisheries and Oceans Canada, by arrangement with the Food and Agriculture Organization of the United Nations.

Smith, C. L., and P. H. Young. 1966. Gonad structure and the reproductive cycle of the Kelp Bass, *Paralabrax clathratus* (Girard), with comments on the relationships of the serranid genus *Paralabrax*. California Fish and Game 52(4):283–292.

Smith, J. L. B. 1961. Fishes of the family Anthiidae from the western Indian Ocean and the Red Sea. Ichthyological Bulletin, Rhodes University, Grahamstown, No. 21:359–369, plates 34 & 35.

Smith, J. L. B. 1965. A rare anthiid fish from Cook Island, Pacific, with a résumé of related species. The Annals & Magazine of Natural History [1964] (Ser. 13) 7(No. 81):533–537, plate 12.

Smith, W. L. 2010. Promoting resolution of the percomorph bush: A reply to Mooi and Gill. Copeia 2010(3):520–524.

Smith, W. L., and M. T. Craig. 2007. Casting the percomorph net widely: The importance of broad taxonomic sampling in the search for the placement of serranid and percid fishes. Copeia 2007(1):35–55.

Smith-Vaniz, W. F., B. B. Collette, and B. E. Luckhurst. 1999. Fishes of Bermuda: History, zoogeography, annotated checklist, and identification keys. American Society of Ichthyologists and Herpetologists Special Publication No. 4. Lawrence, Kansas, pp. i–x + 1–424.

Springer, V. G., and G. D. Johnson. 2004. Study of the dorsal gill-arch musculature of teleostome fishes, with special reference to the Actinopterygii. (Appendix by V. G. Springer and T. M. Orrell: Phylogenetic analysis of 147 families of acanthomorph fishes based primarily on dorsal gill-arch muscles and skeleton.) Bulletin of the Biological Society of Washington, No. 11, pp. i–vi + 1–260, plates 1–205.

Steindachner, F. 1875. Ichthyologische Beiträge [I]. Sitzungsberichte der Akademie der Wissenschaften, Wien. Mathematische - Naturwissenschaftliche Klasse. Abt. 1, 70:375–390, plate 1.

Steindachner, F. 1883. *In*: F. Steindachner and L. Döderlein, Beiträge zur Kenntniss der Fische Japan's. (II.) Denkschriften der Kaiserlichen Akademie der Wissenschaften, Mathematisch-Naturwissenschaftliche Classe, Wien, Abt. 1, 48:1–40, plates 1–7.

Tanaka, S. 1924. Figures and descriptions of the fishes of Japan including Riukiu Islands, Bonin Islands, Formosa, Kurile Islands, Korea, and southern Sakhalin. Vol. 33:607–628, plates 148–150. [In Japanese and English.]

Tanaka, S. 1931. On the distibution of fishes in Japanese waters. Journal of the Faculty of Science, Imperial University of Tokyo, Section 4, Zoology, 3(Pt. 1): 1–90, plates 1–3.

Temminck, C. J., and H. Schlegel. 1843. Pisces. *In*: Fauna Japonica, sive descriptio animalium quae in itinere per Japoniam suscepto annis 1823–1830 collegit, notis observationibus et adumbrationibus illustravit P. F. de Siebold. Parts 2–4:21–72.

Thresher, R. E. 1984. Reproduction in reef fishes. T. F. H. Publications, Inc., Neptune City, New Jersey. 399 pp., many unnumbered illustrations.

Thurman, P., R. McBride, G. D. Dennis, III, and K. J. Sulak. 2004. Age and reproduction in three reef-dwelling serranid fishes of the northeastern Gulf of Mexico Outer Continental Shelf: *Pronotogrammus martinicensis, Hemanthias vivanus & Serranus phoebe* (with preliminary observations on the pomacentrid fish, *Chromis enchrysurus*). USGS Outer Continental Shelf Ecosystem Studies Program Report, USGS Scientific Investigation Report (SIR) 2004–5162 (USGS CEC NEGOM Program Investigation Report No. 2004–03, May 2004).

Tortonese, E. 1986. Serranidae. Pp. 780–792, *in*: P. J. P. Whitehead, M.-L. Bauchot, J.-C. Hureau, J. Nielsen, and E. Tortonese (editors), Fishes of the north-eastern Atlantic and the Mediterranean. Vol. 2. United Nations Educational, Scientific, and Cultural Organization, Paris, pp. 517–1007.

Trunov, I. A. 1976. New species and species recorded for the first time in the pelagic area of the tropical Atlantic of the families Serranidae, Emmelichthyidae and Ariommidae. Journal of Ichthyology, 16(2):229–238. [English translation of Russian in Voprosy Ikhtiologii, 16(2):263–273.]

Tschudi, J. J. von. 1846. Ichthyologie. Pp. ii–xxx + 1–35, plates 1–6, *in*: Scheitlin & Zollikofer, 1844–1846, Untersuchungen über die Fauna Peruana, in 12 parts, St. Gallen.

Valenciennes, A. 1828. *In*: G. Cuvier and A. Valenciennes, Histoire naturelle des poissons. Tome 2. Paris, pp. i–xxi + 3 + 1–490, plates 9–40.

Valenciennes, A. 1833. *In*: G. Cuvier and A. Valenciennes, Histoire naturelle des poissons. Tome 9. Paris, pp. i–xxix + 3 + 1–512, plates 246–279.

Verne, J. 1963. 20,000 leagues under the sea. Reprint of 1869 ed., Airmont Publishing Co., New York, pp. 1–288.

Wade, C. B. 1946. New fishes in the collections of the Allan Hancock Foundation. Allan Hancock Pacific Expeditions 9(8): 215–228, plates 29–32.

Wales, J. H. 1932. An addition to the fish fauna of the United States. Copeia 1932(2):106.

Walford, L. A. 1974. Marine game fishes of the Pacific coast from Alaska to the equator. Reprint of 1937 ed., T. F. H. Publications, Neptune, New Jersey, pp. 1–205, black & white plates 1–32, color plates 33–69 + 1.

Waller, R. A., and W. N. Eschmeyer. 1965. A method for preserving color in biological specimens. BioScience 15(5):361.

Watson, W. 1996. Serranidae: Sea basses. Pp. 876–899, *in*: H. G. Moser (editor), The early stages of fishes in the California Current Region. California Cooperative Oceanic Fisheries Investigations, Atlas No. 33:i–xii + 1–1505.

Weaver, D. C., G. D. Dennis, and K. J. Sulak. 2002. Northeastern Gulf of Mexico Coastal and Marine Ecosystem Program: Community structure and trophic ecology of fishes on the Pinnacles Reef tract. Final Synthesis Report. U. S. Department of the Interior, Geological Survey, Biological Sciences Report, USGS BSR 2001-0008 and Minerals Management Service Gulf of Mexico OCS Region, New Orleans, Louisiana, OCS Study MMS-2002-034, pp. i–xviii + 1–94, app. A–D.

Weaver, D. C., E. L. Hickerson, and G. P. Schmahl. 2006a. Deep reef fish surveys by submersible on Alderdice, McGrail, and Sonnier banks in the northwestern Gulf of Mexico. Pp. 69–87, in: J. C. Taylor (editor), Emerging technologies for reef fisheries research and management. NOAA Professional Paper NMFS 5, pp. 1–116.

Weaver, D. C., D. F. Naar, and B. T. Donahue. 2006b. Deepwater reef fishes and multibeam bathymetry of the Tortugas South Ecological Reserve, Florida Keys National Marine Sanctuary, Florida. Pp. 48–68, in: J. C. Taylor (editor), Emerging technologies for reef fisheries research and management. NOAA Professional Paper NMFS 5, pp. 1–116.

Webb, R. O., and M. J. Kingsford. 1992. Protogynous hermaphroditism in the half-banded sea perch, *Hypoplectrodes maccullochi* (Serranidae). Journal of Fish Biology 40:951–961.

Weber, F. 1801. Observationes entomologicae. Kiliae.

Weber, M. 1913. Die Fische der Siboga-Expedition. E. J. Brill, Leiden, pp. i–xii + 1–710, pls. 1–12.

Wenner, C. A., W. A. Roumillat, and C. W. Waltz. 1986. Contributions to the life history of Black Sea Bass, *Centropristis striata*, off the southeastern United States. Fishery Bulletin 84(3):723–741.

White, W. T. 2011. *Odontanthias randalli* n. sp., a new anthiine fish (Serranidae: Anthiinae) from Indonesia. Zootaxa 3015:21–28.

Whitley, G. P. 1927. Studies in ichthyology. No. 1. Records of the Australian Museum 15:289–304, plates 24 & 25.

Wiley, E. O., and G. D. Johnson. 2010. A teleost classification based on monophyletic groups. Pp. 123–182, in: J. S. Nelson, H.-P. Schultze, and M. V. H. Wilson (editors), Origin and phylogenetic interrelationships of teleosts. Dr. Friedrich Pfeil, München.

Williams, J. T., and R. L. Shipp. 1980. Observations on fishes previously unrecorded or rarely encountered in the northeastern Gulf of Mexico. Northeast Gulf Science 4(1):17–27.

Wu, K.-Y., J. E. Randall, and J.-P. Chen. 2011. Two new species of anthiine fishes of the genus *Plectranthias* (Perciformes: Serranidae) from Taiwan. Zoological Studies 50(2):247–253.

Yáñez-Arancibia, L. A. 1975. Zoogeografía de la fauna ictiólogica de la Isla de Pascua (Easter Island). Anales del Centro de Ciencias del Mar y Limnología de la Universidad Nacional Autónoma de México 2(1):29–51.

Zajonz, U. 2006. *Plectranthias klausewitzi* n. sp. (Teleostei, Perciformes, Serranidae), a new anthiine fish from the deep waters of the southern Red Sea. aqua, International Journal of Ichthyology 12(1):19–26.

TABLES

Table 1. Summary of selected characters in Atlantic and eastern Pacific genera of Anthiinae, based on data from species occurring in those oceans (exceptional values in parentheses).

N/A = not applicable. *Data from only one of two species in the genus.

Characters	Acanthistius	Anatolanthi.	Anthias	Baldwinella	Caprodon
Spines in dorsal fin	11–13	10	10 (11)	10 (11)	10 (11)
Soft rays in dorsal fin	14–18	16	14 or 15 (13, 16)	14 or 15 (13, 16)	19–21
Soft rays in anal fin	7–10	7	(6) 7 (8)	(7) 8 (9)	(7) 8 (9)
Pectoral-fin rays	15–21	21 or 22	16–22	15–21	17 or 18 (16, 19)
Principal caudal-fin rays	9 + 8	8 + 7	8 + 7	8 + 7	9 + 8
Total 1st arch gillrakers	16–23	37	37–48	36–43	33–41
Lateral-line scales	47–70	62 or 63	31–48	36–53	58–71
Caudal-peduncle scales	41–50		16–25	22–29	24–34
Vertebrae	10 + 16	11 + 15	10 + 16	10 + 16	10 + 16
Supraneural bones	3	2	2	2	3
Parapophyses on first caudal vertebra	no	yes	no	no	no
Pleural ribs, pairs	8	9	8	8	8
Epineurals, pairs	9–11	19	11–13	10–13	8 or 9
Antrorse spines on preopercle	yes	no	no	no	no
Scales	ctenoid	ctenoid	ctenoid	ctenoid	ctenoid
Scales with ctenial bases in posterior fields	yes	no	no	no	yes
Secondary squamation	absent	absent	absent	absent	absent
Lateral line interrupted	no	no	no	no	no
Maxilla with scales	yes/no	yes	yes	no	yes
Anterior naris w/filament	no	no	no	no	no
Orbit with fleshy papillae	no	yes	no	no	no
Vomer with teeth	yes	no	yes	yes	yes
Vomerine tooth patch with posterior prolongation	no	N/A	no	no	yes
Tongue with teeth	no	no	yes/no	no	yes

(*continues*)

Table 1. (*continued*)

Characters	*Choranthias*	*Hemanthias*	*Holanthias*	*Hypoplectro.*	*Lepidoperca*
Spines in dorsal fin	(9) 10	10	10	10	10
Soft rays in dorsal fin	(14) 15	14 (12, 13, 15)	15 or 16	19–21 (22)	16–19
Soft rays in anal fin	7 or 8 (9)	(7) 8 (9)	7	(7) 8 (9)	7–9
Pectoral-fin rays	19–22	15–21	19–21	16 or 17 (15, 18)	16 or 17
Principal caudal-fin rays	8 + 7	8 + 7	8 + 7*	9 + 8	9 + 8
Total 1st arch gillrakers	32–39	31–39	38–43	17–20	31–36
Lateral-line scales	46–57	48–71	46–55	48–52 (55)	42–48 (51)
Caudal-peduncle scales	25–28	34–50	24–26*	28–30 (27, 31, 32)	17–19
Vertebrae	10 + 16	10 + 16	10 + 16*	10 + 17	10 + 16
Supraneural bones	2	2	2*	3	3
Parapophyses on first caudal vertebra	yes	yes	no*	no	no
Pleural ribs, pairs	8 (9)	8	8*	8	8
Epineurals, pairs	11 or 12	10–12	11*	9 or 10	11–13
Antrorse spines on preopercle	no	no	no*	yes	no
Scales	ctenoid	ctenoid	ctenoid*	ctenoid	ctenoid
Scales with ctenial bases in posterior fields	no	no	no*	yes	yes
Secondary squamation	absent	absent	present	absent	absent
Lateral line interrupted	usually yes	no	no*	no	no
Maxilla with scales	yes	no	yes	no	yes
Anterior naris w/filament	yes	no	yes/no*	no	no
Orbit with fleshy papillae	no	no	no*	no	no
Vomer with teeth	yes	yes	yes	yes	yes
Vomerine tooth patch with posterior prolongation	no	no	yes	no	no
Tongue with teeth	no	no	yes	no	no

Table 1. (*continued*)

Characters	Meganthias	Odontanth.	Plectranth.	Pronotogra.	Trachypoma
Spines in dorsal fin	10	10	10	10 (11)	12
Soft rays in dorsal fin	17 or 18	15	15 or 16 (17)	15 (13, 14, 16)	13
Soft rays in anal fin	8	7	7	7 (8)	6
Pectoral-fin rays	16 or 17	18	12–17	16–21	17 (18)
Principal caudal-fin rays	8 + 7	8 + 7	9 + 8	8 + 7	9 + 8
Total 1st arch gillrakers	35–39	42 or 43	14–31	34–41	20–23
Lateral-line scales	46–ca. 50	33–38	28–46	35–57	46–56
Caudal-peduncle scales	ca. 25 or 26	16–18	12–22	18–28	35–41
Vertebrae	10 + 16	10 + 16	10 + 16	10 + 16	10 + 16
Supraneural bones	2	2	3	2	3
Parapophyses on first caudal vertebra	no	no	no	no	no
Pleural ribs, pairs	8	8	8	8	8
Epineurals, pairs	11–13	11	12 or 13	10–12	8 or 9
Antrorse spines on preopercle	no	no	yes/no	no	yes
Scales	ctenoid	ctenoid	ctenoid	ctenoid	most cycloid
Scales with ctenial bases in posterior fields	no	no	yes/no	no	yes
Secondary squamation	present	on head	absent	absent	absent
Lateral line interrupted	no	no	no	usually no	no
Maxilla with scales	yes	yes	usually no	yes	no
Anterior naris w/filament	no	no	no	yes	no
Orbit with fleshy papillae	no	no	no	no	no
Vomer with teeth	yes	yes	yes	yes	yes
Vomerine tooth patch with posterior prolongation	no	no	no	usually yes	no
Tongue with teeth	yes	yes	no	usually yes	no

Table 2. Frequency distributions for numbers of soft rays in the dorsal fin of Atlantic and eastern Pacific species of Anthiinae. Counts for *Caprodon longimanus* are from both eastern and western Pacific specimens; those for *Holanthias caudalis* are from the literature.

Species	12	13	14	15	16	17	18	19	20	21	22
Anatolanthias apiomycter					2						
Anthias anthias		1	4	105	4						
Anthias asperilinguis				10							
Anthias cyprinoides				11							
Anthias helenensis				4							
Anthias menezesi				13	1						
Anthias nicholsi			2	46							
Anthias noeli				16	1						
Anthias woodsi			22	1	.						
Baldwinella aureorubens		1	6	84	1						
Baldwinella eos			4	45							
Baldwinella vivanus		4	92								
Caprodon longimanus								17	28	6	
Choranthias salmopunctatus				3							
Choranthias tenuis			2	94		.					
Hemanthias leptus		1	45								
Hemanthias peruanus		2	85	2							
Hemanthias signifer	1	3	66								
Holanthias caudalis				2							
Holanthias fronticinctus				9	1						
Hypoplectrodes semicinctum								6	16	16	1
Lepidoperca coatsii					4	11	1	1			

Table 2. (*continued*)

Species	12	13	14	15	16	17	18	19	20	21	22
Meganthias carpenteri	·					1	1				
Meganthias sp.							1				
Odontanthias hensleyi				4							
Plectranthias exsul				3	3						
Plectranthias garrupellus				9	34	1					
Plectranthias nazcae					5						
Plectranthias parini					2						
Pronotogrammus martinicensis		1	5	69	1						
Pronotogrammus multifasciatus			2	51							
Trachypoma macracanthus		12									

Table 3. Frequency distributions for numbers of soft rays in the anal fin of Atlantic and eastern Pacific species of Anthiinae. Counts for *Caprodon longimanus* are from both eastern and western Pacific specimens; those for *Holanthias caudalis* are from the literature.

Species	6	7	8	9
Anatolanthias apiomycter		2		
Anthias anthias		117		
Anthias asperilinguis		10		
Anthias cyprinoides		11		
Anthias helenensis		4		
Anthias menezesi		14		
Anthias nicholsi	1	48	2	
Anthias noeli	1	16		
Anthias woodsi		22	1	
Baldwinella aureorubens		3	87	3
Baldwinella eos		1	49	1
Baldwinella vivanus			95	1
Caprodon longimanus		1	48	2
Choranthias salmopunctatus		3		
Choranthias tenuis		1	92	3
Hemanthias leptus		1	45	
Hemanthias peruanus		1	85	3
Hemanthias signifer		1	69	
Holanthias caudalis		2		
Holanthias fronticinctus		10		
Hypoplectrodes semicinctum		1	37	1
Lepidoperca coatsii		1	12	4
Meganthias carpenteri			2	
Meganthias sp.			1	
Odontanthias hensleyi		4		
Plectranthias exsul		6		
Plectranthias garrupellus		44		
Plectranthias nazcae		5		
Plectranthias parini		2		
Pronotogrammus martinicensis		75	1	
Pronotogrammus multifasciatus		52	2	
Trachypoma macracanthus	12			

Table 4. Frequency distributions for numbers of rays in the pectoral fin of Atlantic and eastern Pacific species of Anthiinae. Counts for *Caprodon longimanus* are from both eastern and western Pacific specimens; those for *Holanthias caudalis* are from the literature.

Species	12	13	14	15	16	17	18	19	20	21	22
Anatolanthias apiomycter										1	1
Anthias anthias							26	63	17	—	1
Anthias asperilinguis							2	8			
Anthias cyprinoides								3	8		
Anthias helenensis								1	3		
Anthias menezesi						1	10	3			
Anthias nicholsi							3	38	9	1	
Anthias noeli							1	14	2		
Anthias woodsi					1	—	22				
Baldwinella aureorubens				5	37	44	2	2			
Baldwinella eos					1	36	10	1			
Baldwinella vivanus					1	1	23	65	5	1	
Caprodon longimanus					1	19	29	2			
Choranthias salmopunctatus									1	1	
Choranthias tenuis								8	60	10	
Hemanthias leptus						2	32	9	1		
Hemanthias peruanus				1	2	47	35	2			
Hemanthias signifer					1	3	9	51	6	1	
Holanthias caudalis										2	
Holanthias fronticinctus								1	8	1	
Hypoplectrodes semicinctum				1	8	29	1				
Lepidoperca coatsii					8	8					
Meganthias carpenteri					1	1					
Meganthias sp.						1					
Odontanthias hensleyi							4				
Plectranthias exsul					4	1					
Plectranthias garrupellus	2	37	3								
Plectranthias nazcae					4	1					
Plectranthias parini				1	1						
Pronotogrammus martinicensis					5	64	6				
Pronotogrammus multifasciatus								16	27	5	
Trachypoma macracanthus						11	1				

Table 5. Frequency distributions for total numbers of gillrakers on first gill arch in Atlantic and eastern Pacific species of Anthiinae. Counts for *Caprodon longimanus* are presented separately for eastern and western Pacific specimens; those for *Holanthias caudalis* are from the literature.

Species	14	15	16	17	18	19	20	21	22	23	24	25	26	27	28	29	30	31
Hypoplectrodes semicinctum				6	21	8	4											
Plectranthias exsul													1	1	2	2		
Plectranthias garrupellus	3	14	18	6	2													
Plectranthias nazcae															3	1	—	1
Plectranthias parini													1	—	1			
Trachypoma macracanthus							5	3	3	1								

Species	31	32	33	34	35	36	37	38	39	40	41	42	43	44	45	46	47	48
Anatolanthias apiomycter							2											
Anthias anthias								2	7	14	36	12	13	4	1	4		
Anthias asperilinguis								4	3	3								
Anthias cyprinoides								1	6	3	1							
Anthias helenensis											2	1	1					
Anthias menezesi											1	—	1	5	2	2	1	2
Anthias nicholsi									1	8	11	18	9	3				
Anthias noeli							1	6	5	4	1							
Anthias woodsi								8	12	3								

Table 5. (*continued*)

Species	31	32	33	34	35	36	37	38	39	40	41	42	43	44	45	46	47	48
Baldwinella aureorubens						2	9	14	23	26	12	5	1					
Baldwinella eos								2	8	18	15	4	3					
Baldwinella vivanus						5		4	30	18	27	11	5					
C. longimanus east Pacific				6	3	7	1											
C. longimanus west Pacific			1	2	1	5	6	6	6	2	3							
Choranthias salmopunctatus				1	1													
Choranthias tenuis				5	22	30	13	4	2									
Hemanthias leptus				1	3	10	17	10	5									
Hemanthias peruanus	9	38	26	11	2	14	10	2										
Hemanthias signifer			1	12	31													
Holanthias caudalis											1	1						
Holanthias fronticinctus								2	1	1	3	1	2					
Lepidoperca coatsii	1	4	5	3	2	2												
Meganthias carpenteri					1	—	1											
Meganthias sp.									1									
Odontanthias hensleyi												2	2					
Pronotogrammus martinicensis				2	6	10	19	18	11	5	3							
Pronotogrammus multifasciatus						2	4	4	21	12	12							

Table 6. Frequency distributions for numbers of lateral-line scales in Atlantic and eastern Pacific species of Anthiinae. Counts for *Caprodon longimanus* are presented separately for eastern and western Pacific specimens; those for *Holanthias caudalis* are from the literature. One count per specimen except counts from both sides included for *Plectranthias parini* and *Odontanthias hensleyi*.

Species	28	29	30	31	32	33	34	35	36	37	38	39	40	41	42	43	44	45	46	47	48	49	50	51
Anthias anthias										8	13	21	23	14	3	1	1							
Anthias asperilinguis									1	3	4	1	—	1										
Anthias cyprinoides											2	1	2	2	1	1								
Anthias helenensis										1	1	—	—	2										
Anthias menezesi									1	7	—	1	1											
Anthias nicholsi				8	25	9	3																	
Anthias noeli											1	2	2	3	1	4	3	—	1					
Anthias woodsi															2	2	3	9	5	1	1			
Lepidoperca coatsii															1	4	3	2	1	3	2	—		
Meganthias carpenteri																			1	—	—	—	1	
Meganthias sp.																				1				
Odontanthias hensleyi						1	1	1	2	1	1													
Plectranthias exsul													1		1	—	2	1	1					
Plect. garrupellus	10	21																						
Plectranthias nazcae									1	—	—	1	1	1	1									
Plectranthias parini										2	—	1	1											

Table 6. (*continued*)

Species	35	36	37	38	39	40	41	42	43	44	45	46	47	48	49	50	51	52	53	54	55	56	57
Baldwinella aureorubens								2	8	13	24	15	8	5	3	1							3
Baldwinella eos		9	7	10	10																		
Baldwinella vivanus						3				1	5	9	11	15	16	8	5	4	1				
Choranthias salmopunctatus													1	—	1								
Choranthias tenuis																	2	5	11	17	12	5	
Holanthias caudalis												1	1										
Holanthias fronticinctus																4	—	3	2	—	1		
Hypoplectrodes semicinctum														5	8	13	8	3	—		2		
Pronotogrammus martinicensis	1	4	7	16	15	14	3	—	1														
Pronotogrammus multifasciatus												1	2	5	4	6	5	4	3	4	1	2	1
Trachypoma macracanthus												1	—	1	—	2	1	1	—	—	1	1	

Species	48	49	50	51	52	53	54	55	56	57	58	59	60	61	62	63	64	65	66	67	68	69	70	71
Anatol. apiomycter													1	1	1	1	3	2						
C. long. east Pacific											1	1	1	4	2	3	3	3					1	1
C. long. west Pacific														2	3	2	3	3	6	8	2	—		
Hemanthias leptus							2	4	4	8	11	6	4	4	2	1								
Hemanthias peruanus	2	2	6	20	20	23	20	6	5	2	—	1												
Hemanthias signifer													1	2	5	9	8	12	12	10	7	2	—	1

Table 7. Frequency distributions for numbers of circum-caudal-peduncular scales in Atlantic and eastern Pacific species of Anthiinae. Counts for *Caprodon longimanus* are presented separately for eastern and western Pacific specimens.

Species	12	13	14	15	16	17	18	19	20	21	22	23	24	25	26	27	28	29
Anthias anthias							2	2	3	14	11	5						
Anthias asperilinguis						3	6											
Anthias cyprinoides											6	2	2					
Anthias helenensis							1	1										
Anthias menezesi					1	3	5											
Anthias nicholsi						5	27											
Anthias noeli											1	7	5	2				
Anthias woodsi										1	13	2	1					
Baldwinella aureorubens											1	8	16	17	14	2		
Baldwinella eos											5	6	10	2				
Baldwinella vivanus													1	3	14	27	12	12
Choranthias salmopunctatus															1			
Choranthias tenuis														1	5	13	4	
Holanthias fronticinctus													1	3	1			
Lepidoperca coatsii						2	7	2										
Meganthias carpenteri														1	1			
Odontanthias hensleyi					1	1	2											
Plectranthias exsul									3	1	1							

Table 7. (*continued*)

Species	12	13	14	15	16	17	18	19	20	21	22	23	24	25	26	27	28	29
Plectranthias garrupellus	1	—	28															
Plectranthias nazcae							3											
Plectranthias parini					1	1												
Pronotogrammus martinicensis							2	4	13	22	6	1						
Pronotogrammus multifasciatus												5	6	3	3	5	1	

Species	24	25	26	27	28	29	30	31	32	33	34	35	36	37	38	39	40	41
C. longimanus east Pacific	2	—	4	3	2	—	1											
C. longimanus west Pacific							1	2	3	1	1							
Hypoplectrodes semicinctum				1	5	3	4	1	1									
Trachypoma macracanthus												2	1	1	1	—	—	1

Species	34	35	36	37	38	39	40	41	42	43	44	45	46	47	48	49	50
Hemanthias leptus			1	—	—	—	2	6	2	5	2	1	3				
Hemanthias peruanus	4	10	7	4	15	7	7	—	1	2	9	15	8	7	3	1	
Hemanthias signifer			1	—	1	2	—	3	2	7				3		1	2

Table 8. Morphometric data on specimens of three species of *Plectranthias*. Standard lengths are in mm; other measurements in percentages of standard length. Smallest specimen (37.6 mm SL) of *P. nazcae* not included.

(© 2008 Biological Society of Washington. Reproduced from Anderson, 2008, *Proceedings of the Biological Society of Washington*. Reprinted by permission of Allen Press Publishing Services.)

Measurements	*P. nazcae*			*P. exsul*			*P. parini*		
	n	Range	Mean	n	Range	Mean	n	Range	Mean
Standard length	4	115–150	130	6	134–158	147	2	84.7–163	124
Postorbital length of head	4	17.2–19.0	18.3	6	18.3–20.0	19.0	2	19.4–20.1	19.8
Upper jaw, length	4	17.5–18.7	18.2	6	18.3–19.6	19.0	2	18.9–20.6	19.8
Body, depth	4	33.4–36.1	34.4	6	33.7–39.6	36.5	2	35.8–37.9	36.8
Caudal peduncle, depth	4	10.3–10.9	10.7	6	10.4–13.1	11.5	2	11.7–11.9	11.8
Pectoral fin, length	4	30.5–34.3	32.4	5	30.7–33.3	32.1	2	33.9–36.7	35.3
Pelvic fin, length	4	22.3–25.3	23.5	6	23.3–26.1	24.3	2	25.6–28.9	27.2
Anal fin, depressed length	4	26.5–29.6	27.8	5	26.8–30.8	28.8	2	29.0–33.8	31.4
Longest dorsal spine, length	4	16.3–18.2	17.3	6	18.1–19.9	18.9	2	15.3–18.3	16.8
Longest dorsal soft ray, length	4	19.1–23.4	21.9	3	20.6–26.0	23.7	2	26.0–34.5	30.2
Third anal spine, length	4	14.7–15.3	15.0	6	14.7–15.9	15.4	2	15.1–17.9	16.5

MAPS

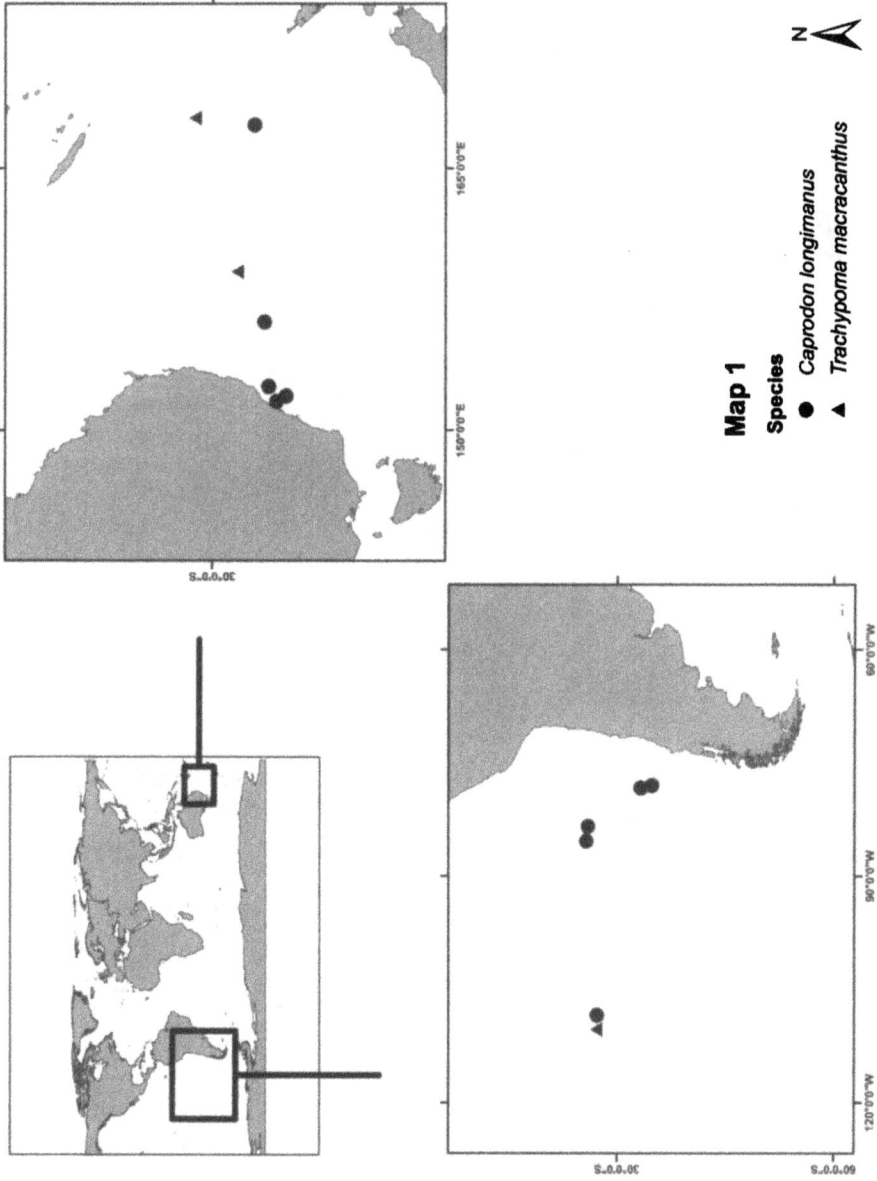

Map 1

Species
- ● *Caprodon longimanus*
- ▲ *Trachypoma macracanthus*

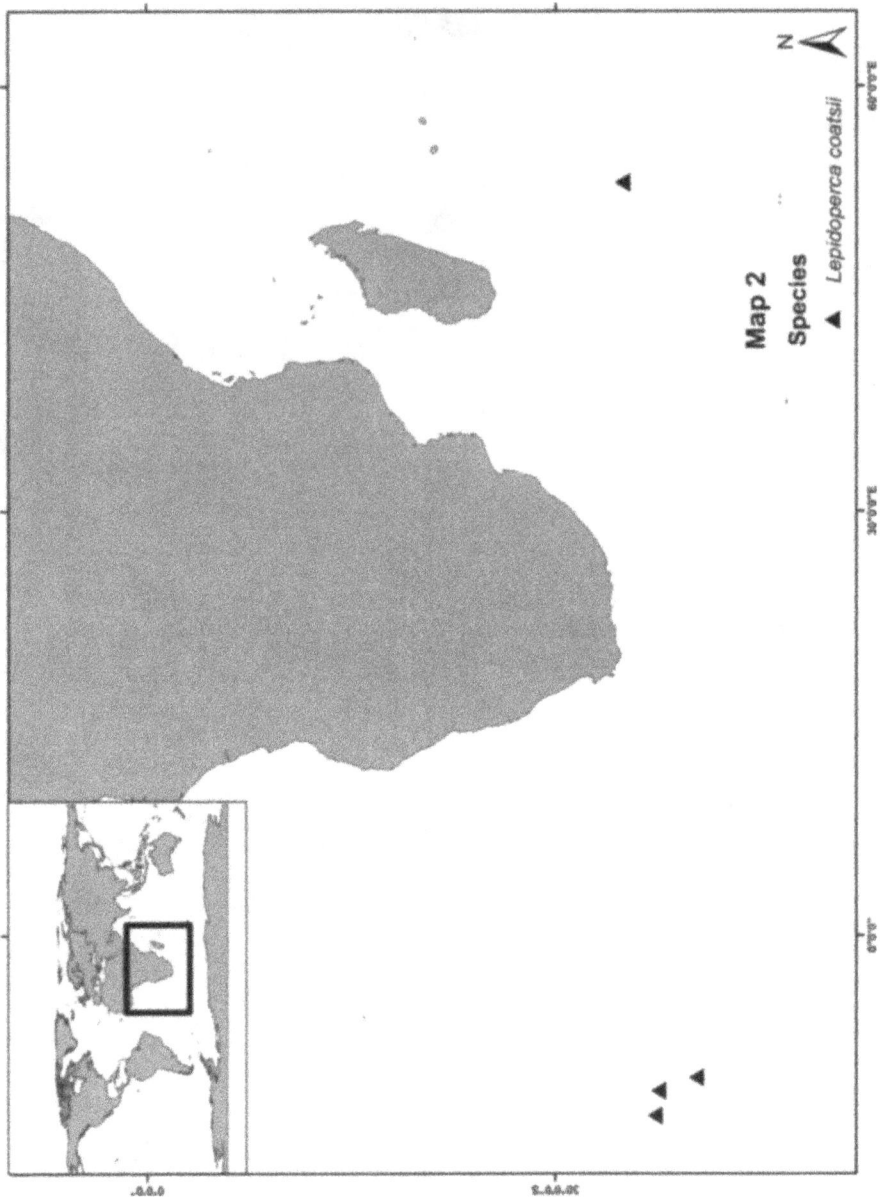

Map 2

Species

▲ *Lepidoperca coatsii*

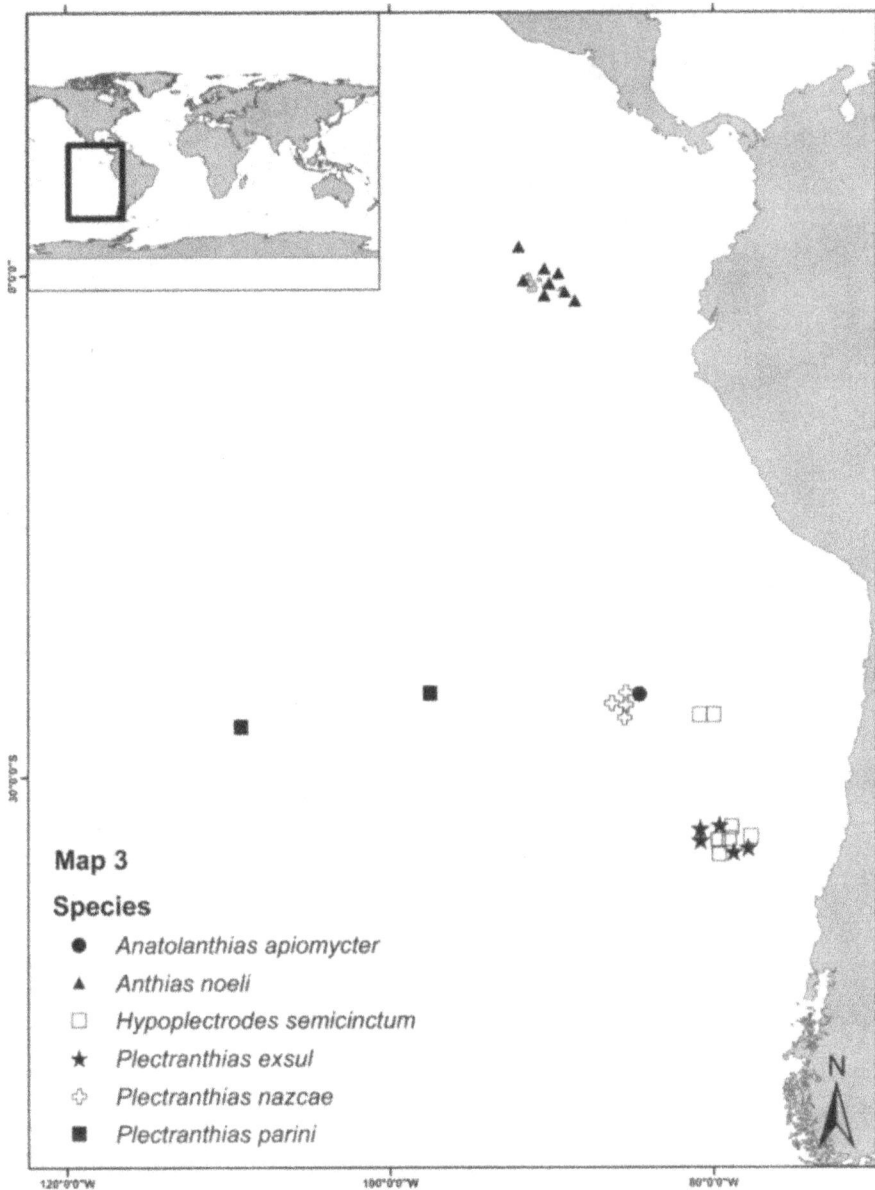

Map 3

Species

- ● *Anatolanthias apiomycter*
- ▲ *Anthias noeli*
- □ *Hypoplectrodes semicinctum*
- ★ *Plectranthias exsul*
- ✢ *Plectranthias nazcae*
- ■ *Plectranthias parini*

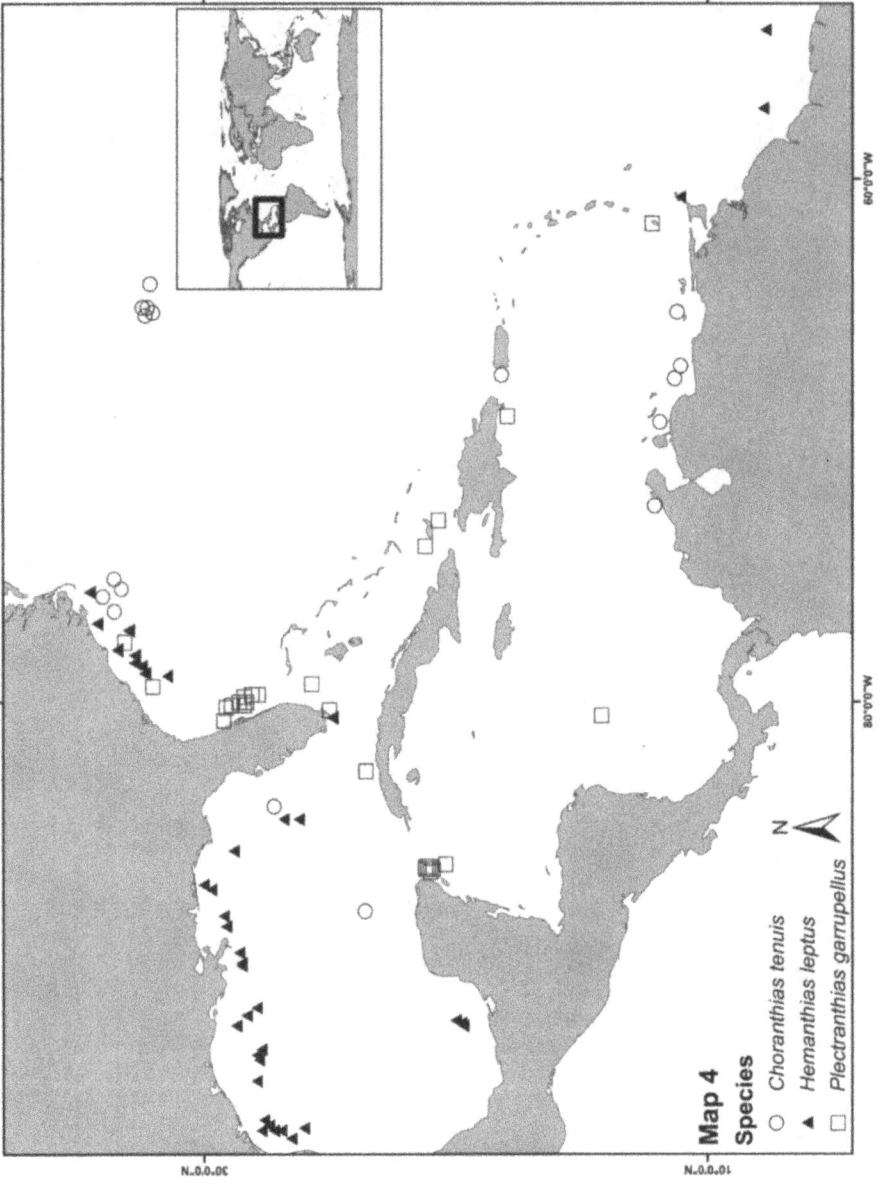

Map 4

Species

○ Choranthias tenuis
▲ Hemanthias leptus
□ Plectranthias garrupellus

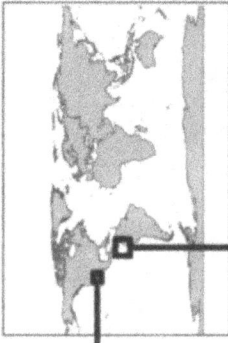

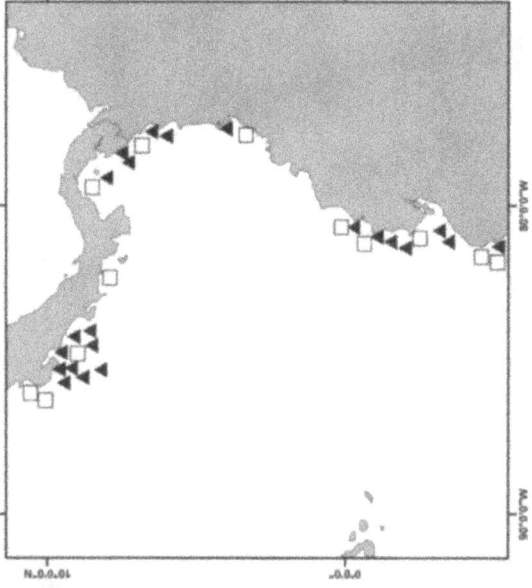

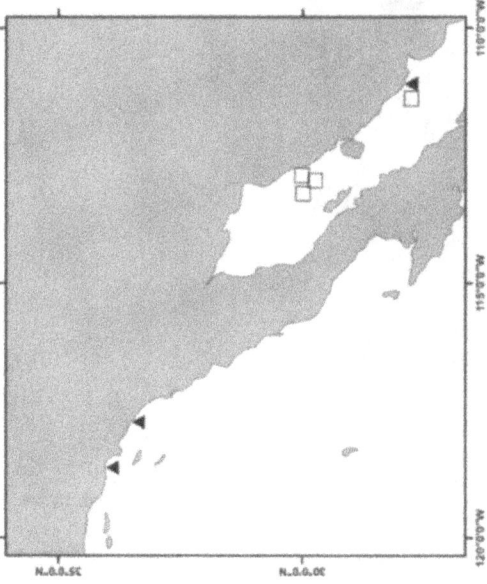

Map 5
Species
□ *Hemanthias peruanus*
▲ *Hemanthias signifer*

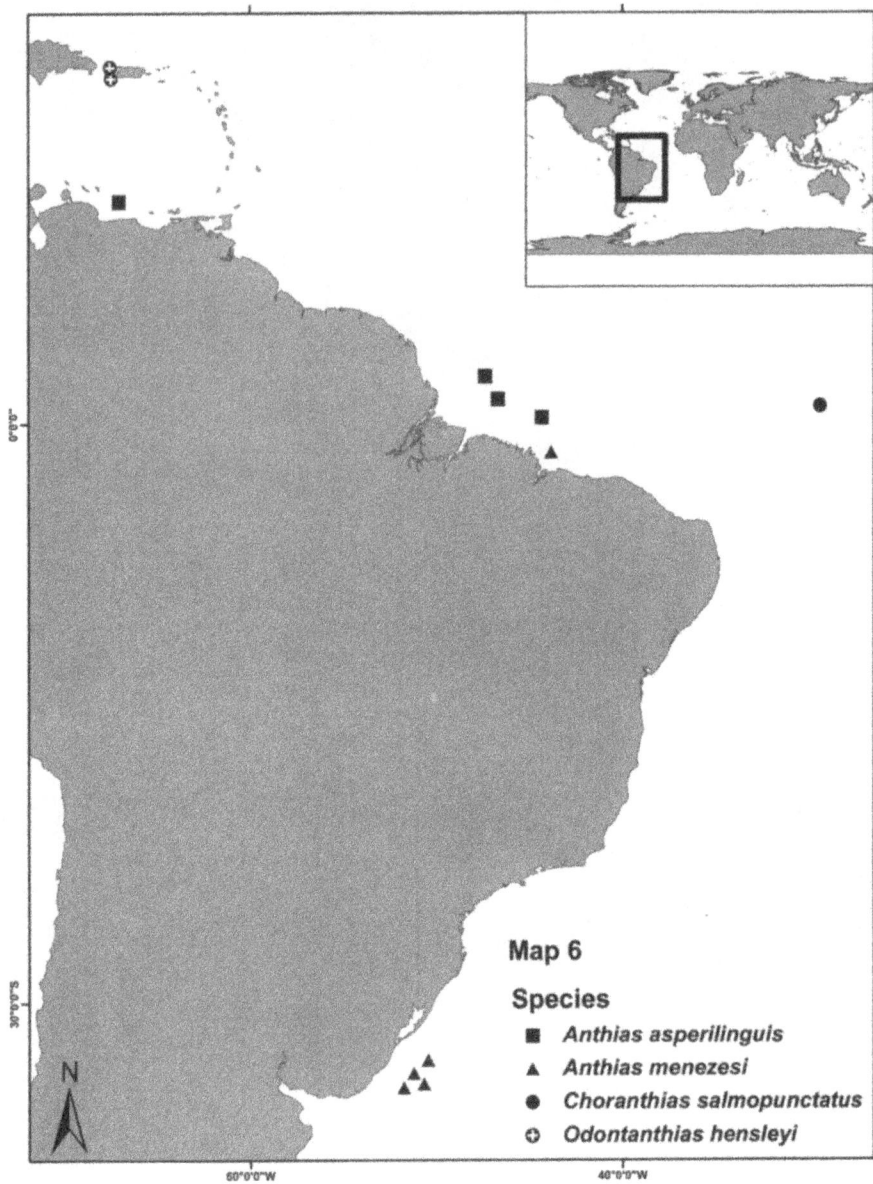

Map 6

Species

■ *Anthias asperilinguis*
▲ *Anthias menezesi*
● *Choranthias salmopunctatus*
⊕ *Odontanthias hensleyi*

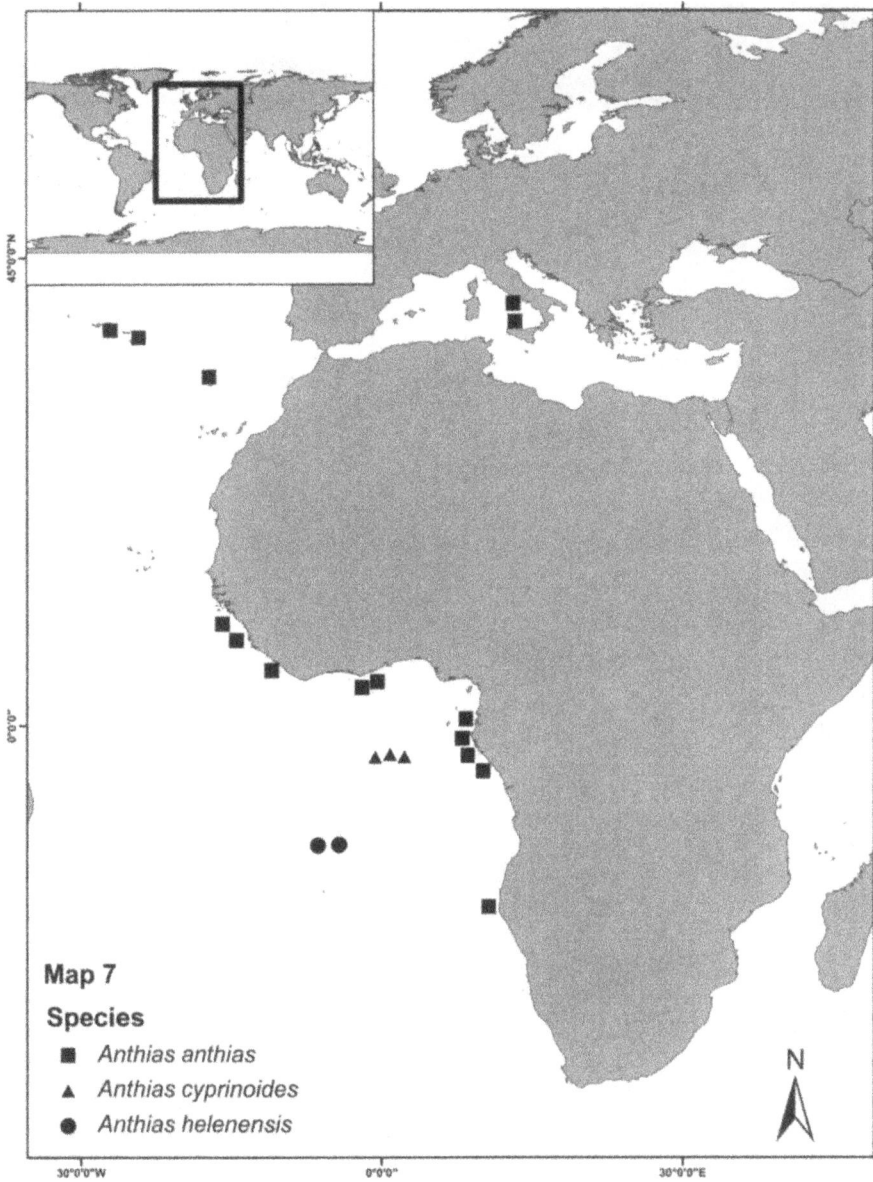

Map 7

Species

■ Anthias anthias
▲ Anthias cyprinoides
● Anthias helenensis

N

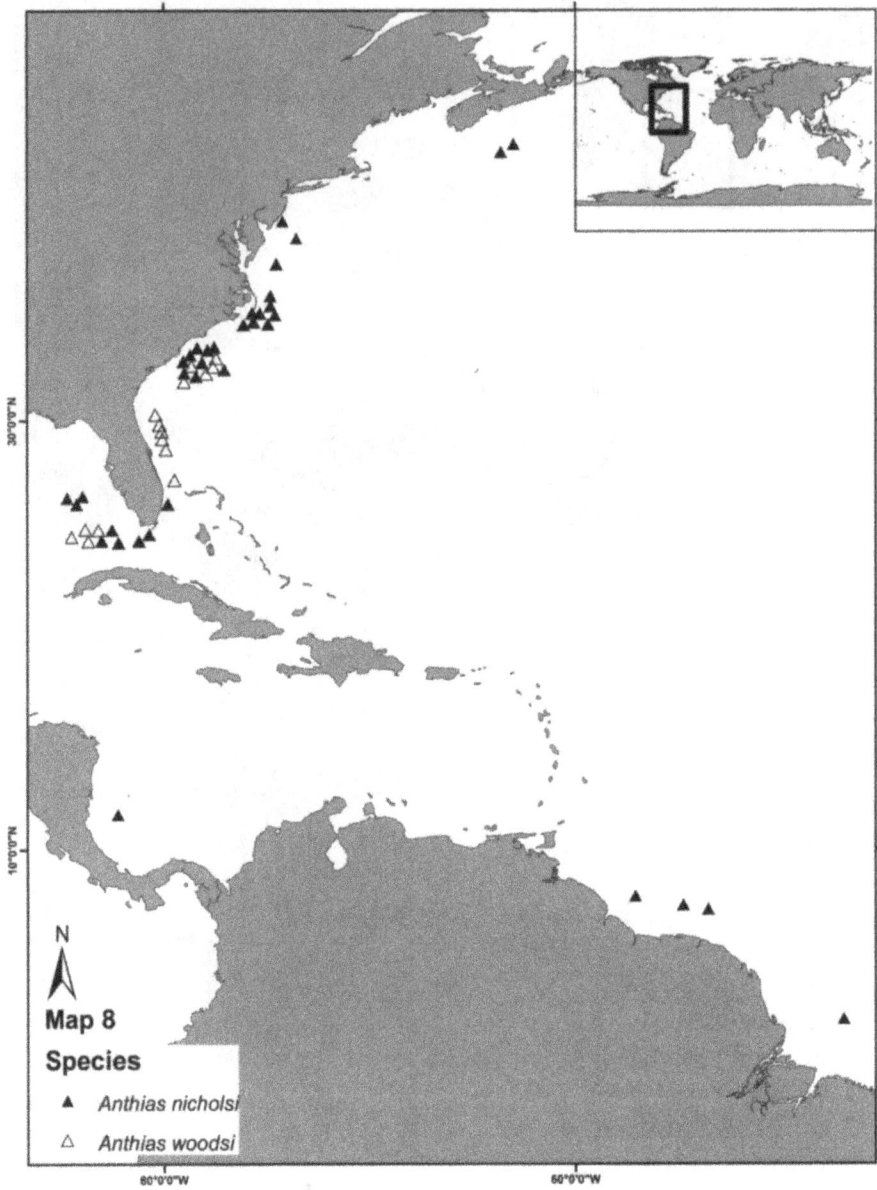

Map 8

Species

▲ Anthias nicholsi

△ Anthias woodsi

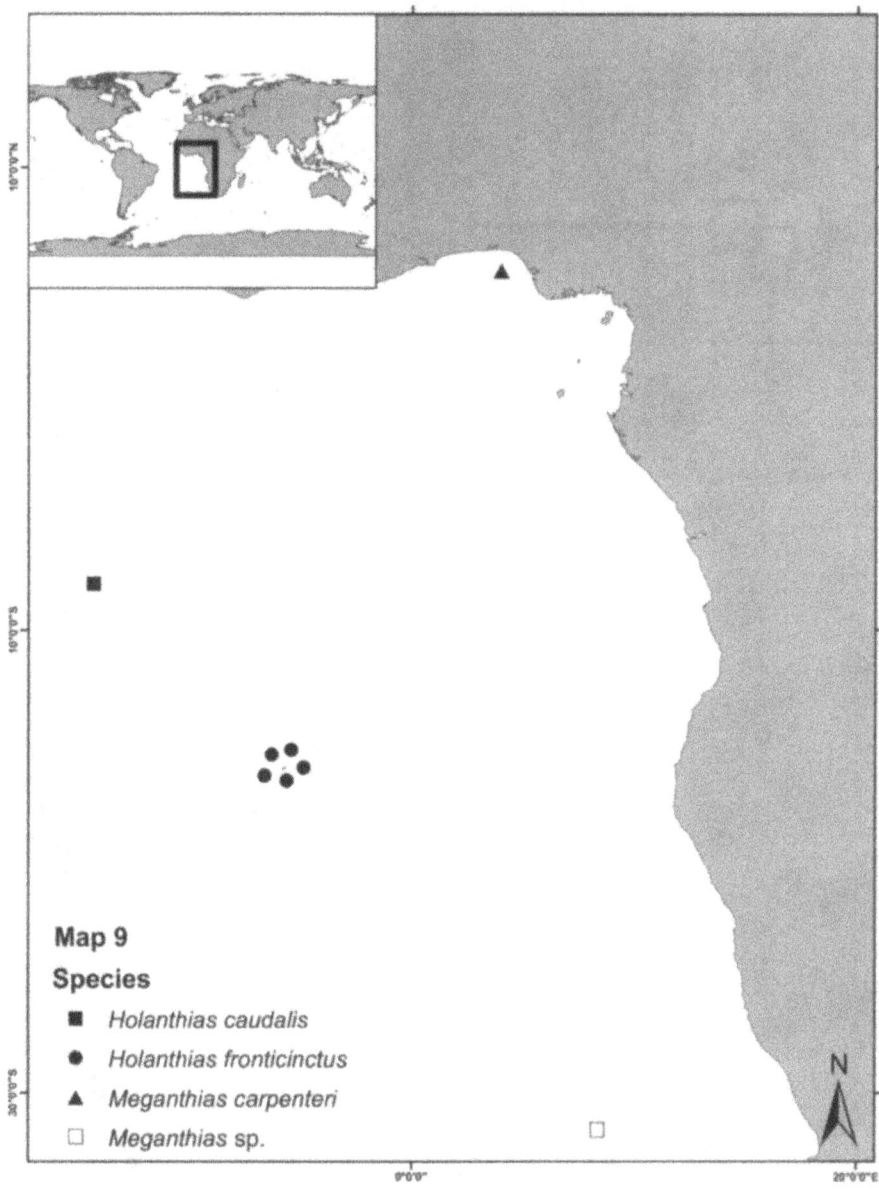

Map 9
Species

■ Holanthias caudalis
● Holanthias fronticinctus
▲ Meganthias carpenteri
□ Meganthias sp.

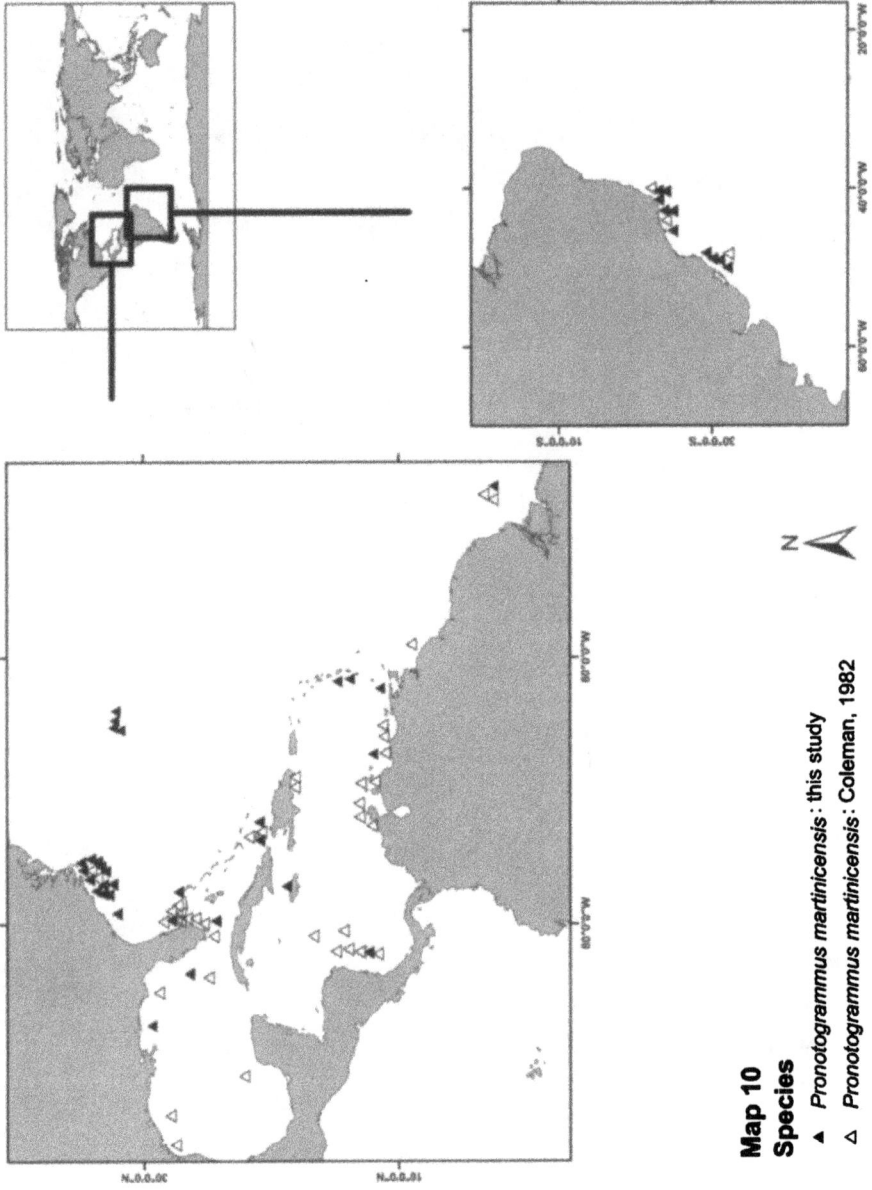

Map 10
Species

▲ *Pronotogrammus martinicensis* : this study

△ *Pronotogrammus martinicensis* : Coleman, 1982

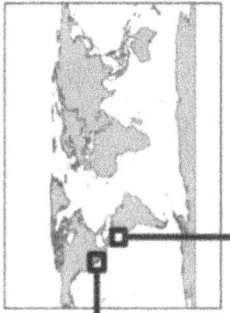

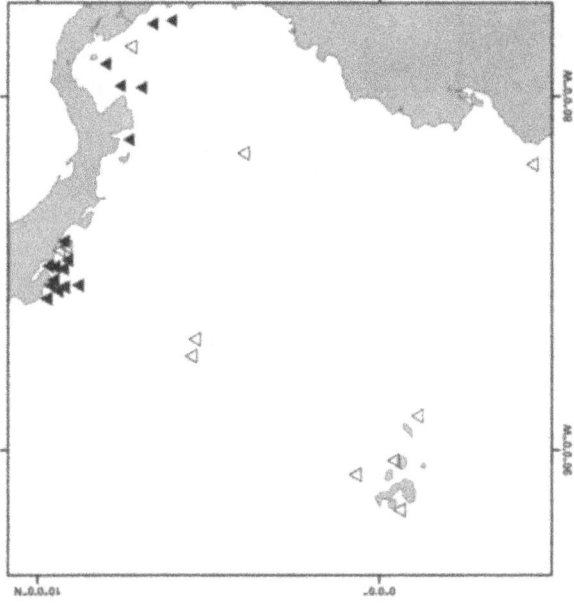

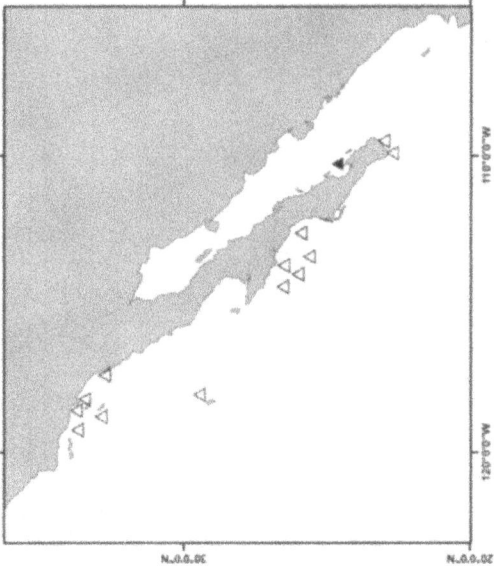

Map 11
Species
▲ *Baldwinella eos*
△ *Pronotogrammus multifasciatus*

N

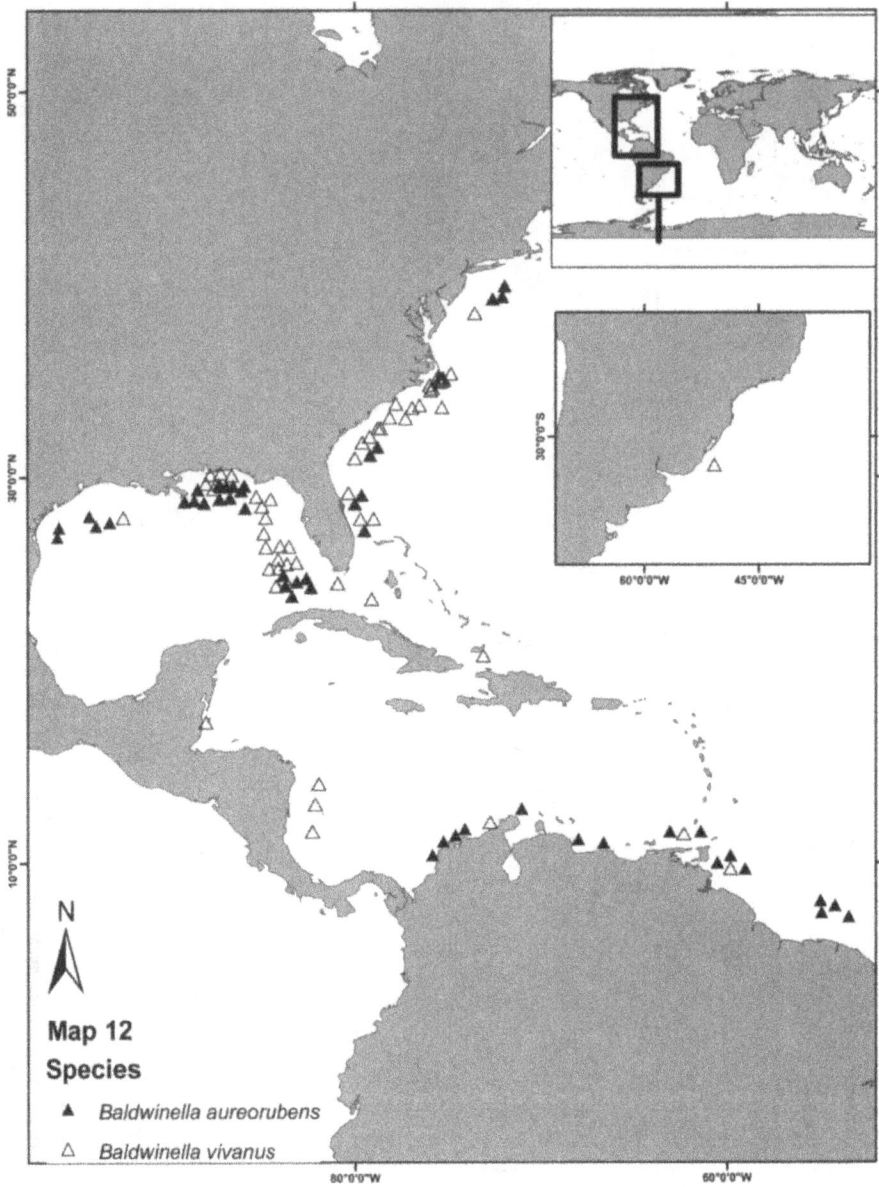

Map 12
Species

▲ *Baldwinella aureorubens*
△ *Baldwinella vivanus*

INDEX

KEYS

Genera, 11
Acanthistius, 13
Anthias, 68
Baldwinella, 110
Choranthias, 61
Hemanthias, 48
Holanthias, 88
Meganthias, 92
Plectranthias, 37
Pronotogrammus, 100

SCIENTIFIC NAMES

Acanthistius, xvii, 3, 5–7, 10, 11, 13, 16
   A. brasilianus, 5, 13
   A. fuscus, 5, 13
   A. ocellatus, 6
   A. patachonicus, 5, 13
   A. pictus, 5, 13
   A. sebastoides, 5, 13, 16
   A. serratus, 5
   Acanthistius sp., 16
Acanthomorpha, 6
Anatolanthias, 7, 8, 12, 59, 60, 107, 119, 120
   A. apiomycter, 8, 12, 59, 60, 107,
      119, 120
Anthia, 67
Anthias, xvii, 1, 2, 7, 8, 12, 23, 25, 26, 35,
      47, 52–54, 60–64, 67–87, 89, 90,
      95, 100, 103, 105–108, 114, 118
   A. anthias, 1, 8, 12, 67–72
   A. asperilinguis, 1, 8, 12, 67, 69, 73, 74
   A. chrysostictus, 95
   A. conspicuus, 8
   A. cyprinoides, 1, 12, 67–69, 74, 75
   A. duplicidentatus, 100, 103
   A. fronticinctus, 87, 89, 90
   A. gordensis, 105–108
   A. helenensis, 1, 68, 75, 76, 89
   A. kelloggi, 35
   A. longimanus, 25, 26
   A. menezesi, 1, 12, 67, 69, 76–78
   A. mundulus, 69
   A. louisi, 100, 105
   A. nicholsi, 1, 8, 12, 67, 68, 78–81

A. noeli, 1, 8, 68, 69, 82–84
A. peruanus, 47, 52–54
A. sacer, 69, 70
A. sacer var. brevipes, 69
A. salmopunctatus, 1, 61, 62, 68
A. schlegelii, 23
A. sechurae, 105, 107
A. squamipinnis, 8
A. tenuis, 1, 2, 60, 63, 64, 68, 100
A. vivanus, 114, 118
A. woodsi, 1, 8, 68, 84–86
Anthias—type 1, 80
Anthias—type 2, 102
Anthias—type 3, 86
Anthiasicus, 47, 48
   A. leptus, 47, 48
Anthiinae, xvii, 1, 5–8, 10, 11, 15, 19, 23,
      29, 35, 47, 54, 59, 60, 67, 87, 91,
      99, 100, 107, 109, 119
   Characters shared, 7
   Feeding habits, 10
   Key to genera, 11
   Major groups, 6
   Maxillary hook, 8
   Otoliths, 7
   Sexuality, 8
Ateleopodidae, 83
Aylopon, 67, 69, 99, 100, 105
   A. algeriensis, 69
   A. canariensis, 69
   A. hispanus, 69
   A. ivicae, 69
   A. martinicensis, 99, 100, 105
   A. nicaeensis, 69
   A. rissoi, 69
Baldwinella, xvii, 1, 2, 7, 8, 11, 103,
      109–117
   B. aureorubens, 110–112
   B. eos, 8, 110, 112–114
   B. vivanus, 2, 8, 103, 110, 113–117
Caesioperca coatsii, 20, 21
Caprodon, 7, 11, 23–28
   C. affinis, 23, 24
   C. longimanus, 11, 23–28
   C. krasyukovae, 23, 24
   C. schlegelii, 23, 24, 26
   C. unicolor, 23, 24, 27

VERNACULAR NAMES

LOCALITIES

www.ingramcontent.com/pod-product-compliance
Lightning Source LLC
Chambersburg PA
CBHW061753260326
41914CB00006B/1093

* 9 7 8 1 6 0 6 1 8 0 2 2 8 *